Suniti Namjoshi was born in India in 1941 where she worked as an officer in the Indian Administrative Service. Subsequently she taught at the University of Toronto for many years. Her poems, fables and satires have been published widely and read in North America, India, Britain and Australia. She now lives and writes with Gillian Hanscombe in a small village in East Devon, England.

## Other books by Suniti Namjoshi

### Verse

*Poems*
*More Poems*
*Cyclone in Pakistan*
*The Jackass and the Lady*
*The Authentic Lie*
*From the Bedside Book of Nightmares*
*Flesh and Paper* (with Gillian Hanscombe)

### Verse and Fables

*St Suniti and the Dragon*
*Feminist Fables*
*The Blue Donkey Fables*
*Because of India*

### Fiction

*Aditi and the One-Eyed Monkey* (children's)
*The Conversations of Cow*
*The Mothers of Maya Diip*

### Translation

*Poems of Govindagraj* (with Sarojini Namjoshi)

# Building
# Babel

## Suniti Namjoshi

Spinifex Press Pty Ltd
504 Queensberry Street
North Melbourne, Vic. 3051
Australia
spinifex@publishaust.net.au
http://www.publishaust.net.au/~spinifex

First published by Spinifex Press, 1996

Typeset in Stempel Garamond by
    Lynne Hamilton and Claire Warren
Cover artwork by Christina Mowle
Cover design by Liz Nicholson, Design Bite
Printed and bound by McPherson's Printing Group

National Library of Australia
Cataloguing-in-Publication data:

Namjoshi, Suniti, 1941 – .
Building Babel.

ISBN 1 875559 56 6

I. Title.

823.914

*For the Reader,*
*that Sweet Barbarian*

# CONTENTS

# INTRODUCTION 1

## AN EXPLANATION

### Crone Kronos, Talking Creatures, Shifting Identities, Building Babel...

## Pantheism

I happened to grow up in India and come from a Hindu background. This means that I took it for granted that the gods are everywhere and in everything – not just the glory of God. This is a useful notion for poets, since it's then perfectly normal for a thing to "glow" or to become "numinous", to bear significance, to become an image that carries a world of meaning. In the Christian tradition poets have to resort to Greek myth, or to the "pathetic fallacy" or to an animistic view of nature. "Idol" is a ridiculous word, which perhaps makes sense from a Judaeo-Christian perspective, but not from a Hindu one. A particular stone is no more "god" and no less so than any other stone. It's a matter of feeling a sense of the sacred. (The word "murti" can't be translated as "idol".)

In the West I suddenly realised one day that "secular" meant "not religious". I had grown up thinking it meant lots of religions and everybody's holidays. For someone with a Hindu background there's probably no such thing as "not religious".

When they were showing *The Mahabharat* in India, people would have a bath, compose their minds, and then sit reverently before their television sets. For that half hour or whatever it was, even the shopkeepers would stop doing business. India wasn't at a standstill exactly, because people are always at a standstill and always scurrying, but the concentration was touching. In my view – in western terms – people were being "literalists of the imagination"; and that is what a fable or a poem requires.

In *Building Babel* The Black Piglet and Solitude keep making up stories about Crone Kronos. Crone Kronos is Time, of course, the mystery. If you like, I invented her; but if you were to ask whether these stories are "sacred" or "secular", I wouldn't be able to answer.


## Heresy And Blasphemy

By "inventing" gods and goddesses or making up stories about them, do I commit heresy? Do I blaspheme? I'm not entirely sure how a Hindu might blaspheme, or what might constitute a Hindu heresy.

You say, "Well, what about Hindu fundamentalism then?" To which I reply, "It doesn't make sense. Hindu fundamentalism is a contradiction in terms."

I sometimes think that it is only by making up stories about the gods that we can engage in the effort to understand, to worship. There is no contradiction between "understanding" and "worshipping". Why should there be? Understanding is worship. The history of ideas is different. India didn't have Cartesian rationalism, Hume's scepticism or Locke's empiricism. *The Upanishads*, which are philosophical texts, are no less and no more important than say the *Gita*, which is embedded in an epic. Every re-telling of a myth is a re-working of it. Every hearing or reading of a myth is a re-creation of it. It is only when we engage with a myth that it resonates, that it becomes charged and re-charged with meaning.

And yet, Hindus must have some sense of outrage about things that are done or said concerning their religion. The closest I can come to it is a sense of disrespect. The Grand Rebellion of 1857 (Indian term) or the The Great Mutiny (British term) is said to have happened because of a rumour that the cartridges the sepoys had to bite contained beef fat (pork fat in the case of Muslims). Or even today if a rumour goes around that a Muslim has committed a "sacrilegious" act – e.g. peed near the image of the god, Ganesh, during the Ganesh Chaturthi festival, which is an important one in Poona, where I come from – then there's every likelihood of the most bloodthirsty riots. What has happened? A god has been insulted? How can a human being insult a god? Poets rant and rave at the gods (Sant Mira Bai, Sant Tukaram, the metaphysical poets, Milton); but the poem is a relationship, the sense of god is a state of mind, and the hammering and the battering is an effort to understand the mystery. It is not a god, who has been insulted, but "good Hindus". In my opinion, when Hindus take up cudgels on behalf of a god, they are not being good Hindus at all, they are being bad poets.

When good Hindus sit before their television sets watching *The Mahabharat* or *The Ramayan*, they are being good listeners or readers of poetry. And a good listener in the act of re-creating is a good poet. That isn't enough. The function of a good poet is to visualise, not fantasise, or as Auden says somewhere in *The Dyer's Hand*, the function of poetry is to "disenchant". As I understand it, this means an acknowledgement of human limitation. We try to visualise the mystery in any way we can, particularly by means of poetry, but the images are of our own making. They are snares, devices, nets, in which we try to catch some sense of the unknown, Crone Kronos, if you like. This means that to insist that a particular story about a god is a "fact" is to "fantasise". It is to insist that a particular snare or device is the definitive one, and that it is something more than a human effort, however inspired. It springs from a desire to constrict "vision" to human need without the awareness that that is what is happening, i.e. without the "disenchantment". (Perhaps that is why the Judaeo Christian religions sometimes insist that a particular text is "divine" utterance i.e. not limited by humanity. To this I would reply, "Perhaps. But the problem is that the listeners, those who re-create the text, are limited by their humanity.")

## Miracles

I've never understood why miracles are considered some sort of "proof" by Hindus, Christians or anybody else. Their chief value is to awaken a sense of the sacred. But the sacred is inherent in everything, it is up to us to release it by our perception. That is what poets and artists do. In that sense they are purveyors of miracles. Wordsworth's daffodils, Blake's sunflower or even poor Solly's re-visionings are as miraculous as an image of Ganesh drinking milk. They mean something. We suddenly see.

## Metaphor

I don't think any poet understands why people use the word, "metaphor" dismissively: e.g. that Jesus was the Son of God is "only" a metaphor, or that heaven and hell are "only" metaphors. A powerful metaphor is the source of revelation. In my view the notion of re-incarnation is a metaphor, and a very powerful one. We become what we would like to be. It's devastating!

The Greek myths have metamorphosis, the Hindu ones often have metempsychosis. The notions aren't identical, but they are related. I was in Greece for a week only once. I liked being there. I felt comfortable with the landscape, and also a little sorry because I realised that the dryads and hamadryads had been superceded. In India the gods are still present. They're everywhere, in everything.

One consequence of this may be that someone with a Hindu background doesn't get particularly self-conscious about making up stories about the gods, or about animals or about anything really. Sometimes in the West, people say to me at readings, "Why do you make up stories about animals? Why not people?" In my heart I'm puzzled. My truthful reply would be, "What's the difference?" I don't say that because I realise this might be regarded as a frivolous quip. But what is the difference between talking people and talking animals? The whole universe is eloquent. It's shouting! This inability to see that there is a qualitative – is that the right word? – difference between humans and animals, even in a story, may be the result of a Hindu background. A

Hindu wouldn't dream of "animating" animals, i.e. putting a soul into them. They have a "soul" already. There's a soul in everything. I've said in *The Blue Donkey Fables* that I write about animals, not people, because I like them better; but to say, "I like them better" is perhaps only another way of saying that they fire my imagination.

## Identity

Perhaps a one-eyed monkey or a blue donkey or even a black piglet doesn't go to a Christian heaven; but if you think about it from a Hindu perspective, a person doesn't go to heaven as the same person either – or get born as the same person. Who you are is just who you happen to be this time around. This notion vastly reduces the importance attached to identity. At the back of your mind you always know that what you're really supposed to do is get rid of identity, not undertake a quest for it. (This has some bad effects as well: e.g. a certain laziness about doing anything about smashing up an abusive caste system). I thought I was playing with the notion of identity and the arbitrary attributes attached to it in *The Conversations of Cow*, not seeking it, though that has been one interpretation. Sometimes I use my own name in a fable. This is not so much because of a terrific concern with who I am, as because it doesn't matter who I am or who it is. (There are also reasons to do with technique.)

I've set down here my awareness of how a Hindu background has entered my writing. This is only what I think. It may be worthwhile though, for all of us to examine how our backgrounds have affected our thinking – or, as they'd say in Babel, to understand how our memes have mutated and merged. The results can be startling. Some of the myths that emerge in *Babel* have resemblances to Greek or Christian myths. That may be because I write in English. They're inherent in the language. It may also be because they're inherent in experience.

In *St Suniti and the Dragon*, Suniti tells her friends, "An ordinary person going on and on about angels and devils . . . is the human condition." She then gets cross because her friends tell her to do something useful, take up a trade. She says that is what she's doing. Now, that's something else I don't understand: this distinction some people make between work and play, where work is considered serious, and play only frivolous. I don't want to go on about all the things I don't understand. I do want to set out the rules of the game. "Come and play", is an invitation. Building Babel is what people do.

*Suniti Namjoshi*
*Winter 1995*

## INTRODUCTION 2

## AN INVITATION

### *The Web as a Medium for Poetry and Dense Text*

## I    Simultaneity and Sequence

A poem on a page of the World Wide Web can have links to other pages as well as internal links, and it can have graphics and sound. But how is this useful or helpful to Writer or Reader? *[The Web is the more mapped and more easily navigable part of the Internet.]*

Poetry wasn't always a poem on a page. That was merely the medium for broadcasting. For hundreds of years and in many different cultures poems were recited.

The sound was integral to the poem.

Good listeners or readers had an instantly available frame of reference. *[This was their RAM.]* Later, good students had a range of reference works. *[This was their hard disk.]*

Today really good readers of poetry hear the

sound of the poem as they read it in their head, and *they have a frame of reference* that makes the resonances of the poem instantly available.

They also ingest the poem as a sequential as well as a simultaneous piece of work. *The connections between different bits of the poem* aren't just a matter of contiguity. There are links between different sections. And these are held together simultaneously in the head.

Good readers also literally *see the poem's images* in their head.

But most people are not good readers of poetry. In part the problem arises because there is no longer a common frame of reference.

So how can the use of an HTML page help? [HTML or hypertext markup language is a way of setting out a text that allows for links to other bits of text, graphics and sound. *It isn't difficult.*]

Some answers. An HTML page can:
1. Make the images instantly available (graphics).
2. Make the sound instantly available.
3. Make useful information (i.e. the missing frame of reference) instantly available.

*[The links are only almost instantaneous at best, not genuinely instantaneous.]*

These three possibilities just use the technology as a teaching tool, a crutch. They do nothing to the nature of poetry itself. But poetry was never made out of the printed page or a computer screen. Its medium is words, thought. There is, of course, a terrible confusion about the word "medium".

A teaching tool is boring. People don't like being taught. Well, there it is.

## II   Interactivity

A dense text makes huge demands on its reader. It demands that the images be recreated. It demands that sound be heard. It demands that the patterns between the sounds and the images be perceived. It demands that the concentrated references flower in the reader's head. And it demands this on its own terms; it demands that the reader understand and accept the rules in accordance with which the game is being played.

*Let's imagine a Power Struggle between Writer and Reader.*

"Whose terms? Yours or mine? Why should you be the Writer? Why can't I participate?" Reader might ask.

Writer says, "I didn't make up the rules. They are part of the cultural heritage. And where I've changed them, I've written the changes into the text."

Reader sulks. "You've done it all. What's left for me to do? I first have to learn the contents of your goddamn head and your difficult rules, and then swallow your poem."

Writer demurs, "There's everything left for you to do. At least fifty per cent. All that work is interactive. In the end you recreate the poem in accordance with the contents of your own head. It's not win or lose. You get your individual version of the poem. It sings in your brain. And out of that poem you can, if you like, write another poem. There is no power struggle. Nobody wins or loses. I am not trying to fight you. I am trying to please you."

Reader scowls, "I am a Yob of my time. If nobody wins or loses, this game is boring. I prefer football. Besides, there is a power struggle. You are forcing me to think in your terms. You want me to learn your language. Why should I?"

Writer replies, "Because you want to be interactive, and I'm offering you loads of things to do."

For Writer the power struggle amounts to coaxing, cajoling, cudgelling and conning the Reader into playing the game.

For Reader the power struggle is located elsewhere. It begins with the question, "Is this a game I want to play?" Saying "No", is an exercise of power. (The quick switching of channels on TV is part of the attraction of TV over books. The sequentiality of a book offers less choice. Surfing the Net in this respect is like switching channels. You don't have to look at anything for long.)

The next exercise of power occurs when Reader says, "Okay, I'll participate, but on my terms. I'll identify with one side or the other – e.g. Arsenal for football – and play to win or lose by proxy."

"Fine," says Writer, "I'll create a hero or heroine –"

"No. No females," says the Yob, "as hero. Can't identify with a female."

Writer ignores this. " – and you identify with her, and win or lose by proxy. I can even guarantee a win, if you like."

"No, I want to intervene in the game," says Reader. "I want interaction. Not just your boring old text that I'm supposed to accept. I'll play a computer game instead."

"In fact," continues Reader, "I want to be like you. I want to be paid for playing. Set up a quiz, a prize, a game show. If I press the right button, I'll win. Me, personally."

"Look," says Writer, "I am offering you the chance to be the writer. When you read my poem properly, you understand the rules. At least fifty per cent of the poem, you make. And afterwards you can even write your own poem based on the shards of my poem."

"Damn your poems!" yells Reader. "Maximum gain for minimum effort. That's my creed. I don't want to play your game for the sake of playing the game. I want other things out of it too."

"What?"

"1. Being with it. Impressing other people.

2. The delusion of being in control, having power.

3. Money."

"Knowledge is power," offers Writer. "Self knowledge."

"I want," says Reader, "easy power. Easy money. And some understanding on the part of my mates and peers that I have got it."

The point of all this is that computer technology and TV technology have catered to these demands far better than paper and print. Everyday TV corresponds to magazines. Great movies are perhaps "books". The equivalent of books and great movies hasn't been made for the new computer technology yet. Perhaps it will always remain just that – a medium for storing and broadcasting the creations of poets, painters, musicians and movie makers.

*[A floppy disk or CD-ROM can have links as well, but all these links will be confined to the contents of the disk, they won't be links to other pages of the World Wide Web.]*

Can the technology offer the kind of "Interactivity" that Writer is talking about rather than the kind Reader might prefer?

1. Used as a teaching tool it can make the Writer's game easier and faster to understand.
2. It can perhaps offer links within the text to make the patterns visible.
3. It can offer limited options to the Reader, such as "Click here", as well as the more wide ranging and ultimate option "Write a poem".

A computer program is a series of commands. "Click this button" is a command to set a command in operation. Perhaps this knowledge – "I, Reader, am making something happen – even if it's only making an image appear on the page" – provides Reader with some sense of having control.

There is something else Reader can do, if he/she wants the feeling of being Writer and that is STEAL. This is not as shocking as it sounds. All writers steal. The point is that the new technology makes theft easy. The real point is that, therefore, the new technology could make writing a poem or adding to the text more possible, more easy, for Reader.

## III  Collage

> *Nor is there singing school but studying*
> *Monuments of its own magnificence.*
> (Yeats, *Sailing to Byzantium*)

All writing is made of other writing. Words are common currency. They are taken out of the common hoard. Now on the Internet it's easy to make a beautiful page by stealing fine feathers from other pages. Download a copy of a painting by Kandinsky, a tune from somewhere, and jot down a set of favourite links. Put them all together. Maybe it works, maybe not.

The original offer from Writer: "Out of the shards of my poem, write your own poem", is made very easy to do.

Then says Reader, "Okay, but you've published your poem. You're shoving it under my nose. I want the power to broadcast."

Again, on the Internet, this is easy. Reader can set up his or her own Home Page, and anyone in the world who wants to look at this Home Page, can access it for the price of a phone call, or, if they're working in a university, not even that. Then Reader can say, "I am broadcasting to the world. Not just my vote, but my voice, my VOICE is being heard."

## IV  Broadcasting

There's a bit of delusion and a bit of truth here, just as in the matter of being commanded to give a command. "I am broadcasting." Yes, that's true. "But is anyone tuning in?" – That's a problem.

There are differences between the old broadcasting technologies – often called media – and the new one.

1. Downloading someone else's work is much cheaper and easier than xeroxing, or recording a radio program, or making a copy of a TV program.

2. Everyone on the Net, is not just a receiver, but

quite easily a broadcaster as well. The average radio listener doesn't have a transmitter. (Anyone with something as simple as email has just that, and with any luck an email message can cost less than a postage stamp.)

The problem – "Is anyone listening to my broadcast?" – has to do with the interaction between global capitalism and the new technology. Cyberspace is rapidly turning into something very like the real world. Questions of private property arise.

1. Some sites are more worthwhile than other sites. A shop on the high street is better than a shop in the wilderness. Known sites with lots of links or roads to them are more likely to be visited than obscure, isolated ones.

2. Then there's the question of payment: "Should you be paying me for tasting my wares? Or should I be paying you for swallowing my information?" There's the same muddle here as in the real world about the distinctions between advertising, information and entertainment. This is related to publicity. The Reader says, "I'll read a book I've heard about." This has to do with publicity. Or: "I'll visit a site I know about." Publicity again.

3. And what about copyright? Saying, "All Writers steal" is all very well. Writer says, "There are limits. Besides, I, the Writer, want to be paid for my work."

So, "I can make my voice heard and be Writer" is true only to a limited extent. At this point in the power struggle Reader becomes Writer and realises that Writer herself/himself is relatively powerless. I/You/We can shout. Whether we'll be heard is a different matter.

It is at this point then that Writer/Reader engages with the Publisher, who, of course, has problems of her own. It is at this point also that I discard masks; and, as myself, as Suniti, I say to Susan and Renate of Spinifex, "Please, please I want the last chapter of *Building Babel* put on the World Wide Web, with an invitation to Reader to contribute to it. Please. It's the logical conclusion to my book, which is about the process of building culture in the teeth of Crone Kronos. It's aesthetically right, and what's more it's cost effective!"

Susan smiles. Renate keeps faith that I haven't gone mad. "And who decides what is accepted?" Susan asks.

"You do," I reply promptly, "once the book is published. And before that, I do. I've already got some things – just as an encouragement to other Readers."

"Not every Reader's contribution can be accepted," Renate points out gently.

I nod. I agree. "There's an imbalance of power. We decide. But dammit all, Renate," I say earnestly, "we're mortal. We're going to die. Whether we want to or not we'll have to relinquish power. Just like Cinders. Just like Rap Rap. We cannot forever control Babel."

I take a deep breath. I'm going to make a speech. They're patient. "We're in our fifties – well, most of us are. The feminism that we fought for has mutated into strange shapes. Some we don't like. Some we don't recognise. But that's how it is. That's how it must be. Memes mutate. Can we not, may we not, leave a legacy to which others may contribute? True, it will be broken, altered, changed. It will be the broken jewellery in Mad Med's head. But we operate under the aegis of Crone Kronos, and we are all complicit!"

Susan has decided that someone has got to be sensible. She asks, "In practical terms, how will this work?"

I'm ready for that one. I've thought it through. Well, I've thought some of it through. "Like this," I say. "The book is published. Then the last chapter, "The Reader's Text", goes on your Home Page on the World Wide Web with an invitation to Reader to contribute to the memes of Babel. Such contributions as you accept can go on the Home Page. If there are enough, they can constitute an anthology, *Building Babel '97*, *Building Babel '98*. The invitation can also go on the printed page. Eventually, the original text might be forgotten. Don't you see, all poems, all works of art, all patterned thoughts, are really a snare for Crone Kronos!"

Renate smiles, "And so Babel is – ?"

"The Great Barrier Reef!" I tell her recklessly.

"But won't the reef get built anyway? Built and dissolved, and perhaps built over again?"

"Yes," I reply, "but consciousness is everything. Consciousness matters. Let's do it for the hell of it. Let's do it because Crone Kronos has let us have our day!"

And they say, "Yes".


## Invitation from Spinifex

So Suniti, so dear Reader, here's your chance, a chance to contribute to the Building of Babel. The architectural plans are in your head, perhaps you want to collaborate with a friend, draw up a collective design, or put the idiosyncrasies of your head on the Net.

Whichever way you choose to do it, we'd like to hear from you. As Suniti has pointed out it won't be possible to show everything. And your plans will have to come in a format we can handle and will have to conform to our usual rules for manuscript assessment.

Publisher tilts back in her chair, expressing power momentarily through body language. "I don't know about *Building Babel '97, Building Babel '98*," pausing. "We won't know until we see what comes. I can't give guarantees. What I can say is that we will apply the same selection criteria to these contributions as we would to any unsolicited material."

She continues, "We also won't be offering payment for work included on the Home Page. Although, as a writer myself, I'm in favour of it in principle, it isn't practical if we want to continue existing as a small and independent publisher. The Reader/Writer will need to decide whether payment is more important than visibility. Of course, anyone can make their own Home Page and self publish. Virginia Woolf did it. Lots of writers do."

So Suniti, so dear Reader . . . send your contributions to us.

**Text documents should be sent as email to our email address: spinifex@publishaust.net.au For visual and other formats send, preferably, in jpeg.**

For some ideas on what you might send, visit the Babel Building Site on

<http://www.publishaust.net.au/~spinifex/babel.html>

or turn to The Reader's Text at the end of the book.

# PROLOGUE / EPILOGUE

These are the characters. They are pilgrims all. The Wife of Bath is like Little Red. She is Little Red. She is fat and sweaty, her eyes squint. She likes a bit of male flesh, a bit of red meat, a glass of beer. But the good wife is upwardly mobile. She has had an education. She has had opportunity. She has had six sons, a solitary daughter. (The Wife of Bath – had she any children? – She had them all right, but in that bit of life they were not relevant. The Wife knew – when she liked – how *not* to have children – ahead of her time.) Little Red is less raucous, more seemly. Like her ancestress she's an excellent business-woman. Unlike the Wife she can consort with Cinders. It's money that does it. She means well. She has done well. At this point in her life, she would like to do something, build a little something, with *some* of her money. She lives on a ranch in America, a station in Australia – always a supply of good red meat. She also lives in a city in India. There, of course, she is vegetarian, but still a good cook; and on the sly, sometimes, there's the chance of a laugh, a smutty joke, a nod and a wink. How describe Little Red? Blonde or brown, or possibly red headed, but that's not the point. Little Red is ripe and rich, and she is fleshly.

Sister Sol, ah sweet Sister Sol, on the other hand, is skeletal. The wind will sing in her bones, birds will nest in her rib cage – or so she imagines. Such loneliness begets hollowness. Lucky for her she met The Black Piglet, someone or something she could latch on to: not too real and therefore frightening. Someone with whom she could genuinely converse – in the elliptical mode – her preferred style. She's severe, savage, and suggestible, like the sand, the sun, and the wind. All by herself she'd have made nothing. She lives in the desert, of course, always in the desert – by temperament and will power a natural exile. Her Past is her shadow. She carries her Past in order to proclaim she is not who she was. All that she has left behind. Who was she then? A pauper princess, from Russia, Czechoslovakia, or wherever; a high flying executive who lost all his money; and, well, possibly, an unrecognised genius who took to drink. She is, oh, she is like an Olivia, who lacks common sense, both proud and needy. Sweet Sister Sol, when she found her motorbike, adopted a New Look. "At last," she thought, "at last, I am with it!" But she wasn't: habit dies hard. Still, she wasn't as unique as she thought she might be. Sweet Sister Sols, by the dozens and millions, lurking in the dark and hollow places of the human heart, make, perhaps for a kind of sisterhood, but not one she'd relish. Capable, though of loving The Black Piglet with an intensity of passion that is sisterly.

Who is The Black Piglet? A succulent porker. A voyager, a wanderer, a thinker. The Black Piglet liked to tell stories. That was her thing. So many stories infest her head that, as she wanders through the mazes of these multiple histories, she finds herself changing. Her temperament is such that mirrors miss her. She wants to know: What happened next? What did The Piglet do? Or the prince or the princess? And all the while, there she is, with her eyes half shut, under the shade of some friendly tree. She could not have told you which tree it was, until and unless she shut her eyes. A dreamer, a contemplative, not exactly lazy. The granite of the building blocks, the weight and rough surface which made her bleed, was the gift of The Sisterhood; and though difficult, it was in some sense a genuine gift. What is her provenance? Where did she live? She lived in India and told stories, she lived in America and studied physics, she lived in Canada and sculpted ice. She lived in the Badlands and Fatlands. She lived and she died. Chiefly, she lived inside her head and quite enjoyed it. Sometimes she invited other people in. The contents of her head looked like a junk shop, though she had hoped to make a garden in it.

Rap Rap's appearance is well known. Her image

was always for presentation. She is tall and well dressed, but not too well dressed. She wears only a trace of lipstick. Her clothes: a dash of individuality, a modicum of conventionality, and a touch of fashion – nothing excessive. She was dean of her department; she chaired the local council; she was often on the board of several charities. She tended to chair whatever she did. She was able, ambitious, she achieved. She occasionally regrets that she hadn't sketched a little more; but she did sketch a little and she hopes that, one day, her sketches too will be recognised. She admires Cinders. This admiration she keeps to herself. A secret love? Hardly that. Her sex life resembles the clothes she wears. She disapproves of passion. She values friendship.

Cinders, herself, understands this. She smiles on everyone equally. She too has a need for friendship. Her sudden leap from rags to riches was so very sudden that she lost ambition. The prince had it all, and she had the prince. She understood clearly what was required of her: she had to be gracious whatever she did. This she is. Her presence sheds glamour on everything.

Of the minor characters there's little to be said. Snow White and Rose Green are college students, long hair, short skirts, whatever the style. "Me too," they might say, "us too," as they dream their casual dreams. The cat, like the others is an arrant egoist. In his male manifestations he felt excluded. This pleased him sometimes, and sometimes he very much resented it.

He liked Alice. Why, it's hard to say. Perhaps he felt he understood her. In her presence he experienced familiarity. Alice, herself, liked power, and unlike Rap Rap she had no misgivings, no insistent self-image of a calm, benign, well-meaning being. For her, "Who's to be boss?" was the basic question. She had no trouble answering it. She liked control and was perfectly able to exercise it. She liked order and felt perfectly willing to see to it. She has been photographed, illustrated, annotated, and copiously celebrated. "History matters", she is inclined to say, and has given herself up to making it. Her personal life is a little dim, but that no doubt is as it should be.

Then there's the Villain-Villainess: the creature who had nothing and therefore wanted everyone's anything: Envy Incarnate, Malice made moody, which is rubbish, of course. Of the lot of them, Lady Shy was/is always the most beautiful. Beauty bred in the bone, as a poet might say. Sometimes she knew this and sometimes she didn't. This troubled her. A minor weakness. She was in the long run fairly successful. She is not the poor, little rich girl. She's the moderately poor, moderately rich girl, who always comes second – undeservedly. This is galling. She gets no attention, and no sympathy. This too is galling. "Self-image," Cinders whispers, "proper presentation of the surface self." Shy tries, but somehow she never quite gets the shade, the tone, the accent or angle – whatever it takes – exactly right. Bad PR, or bad karma. People disliked her, or rather they disliked what they saw in her mirroring eyes. And yet, once in a while, when she actually understood something, she was able to be extremely kind.

The Three Gorgons, Mad Med, Verity and Charity are Primal Forces. But who could be afraid, overawed, threatened, or even impressed by Sister Charity? There she is, sweet, sickly and apparently unkillable. Who is she then, and where does she live? Oh she's the bloody neighbour who made you a cuppa, even when tired and short of everything. And Verity, that humbug? How could she possibly be a Primal Force? By the power of her name. Her name is almighty. Therefore, despite her foolishness, she is not ineffective. As for Mad Med, do not ask. She's all blood and guts. She's Caliban. She's a baby. A teen-ager. A rowdy. She has an appetite. She lives where she can. She's a squatter. She lives in squats with other squatters, her friends and companions, the other rowdies. She knows she's supposed to be the juvenile lead. Sometimes she practises being girlish, and sometimes she *is* the juvenile lead. Very confusing. Her head is a treasure house of broken jewellery.

Last but not least there's Crone Kronos, together with her siblings, Death and Mme Mem. Obviously, they inhabit the heads of the others, and yours, and mine. They live on the outside and in the inside. Crone Kronos is ubiquitous. Praise be!

The disciples of Crone Kronos build Babel together. Is the hodge podge of their needs reflected in their lives? What do they smell, eat, drink? What are their gardens like? Do they have gardens? Do they make love? Yes, some of them have gardens. And yes, they make love, but not in any sort of saturnalian orgy. Well, not often. And yes, there's the occasional lesbian liaison, but nothing scandalous or particularly spicy. Their private lives – soap opera stuff. Any soap opera. They have children, they have lovers. They get miffed. Oh and they suffer sudden mischance, death and disease. It would be more magnificent perhaps not to be a member of just any soap opera. A change of name, a change of genre, would do the trick; but repetition happens. That's how it is. Be gentle if you will. They are not dolls, they are live flesh, with chaotic suns and worlds flaring within.

Where precisely was Babel built? In the Gobi or the Sahara? Or the Rajputana Desert? Where do the sands sweep to the sea? Babel was built in your brains cells. Surely you know, the memes of Babel are colonists. They are your RAM, your instantly available, accessible memory. The ruins of Babel, the growth and degradation, the endless adaptation, the building and rebuilding, they are on your hard disk.

# I

## *PIECE FOR SOLOISTS: WHAT THE SISTERS SAID*

## I    IF TWO SISTERS

If two sisters sat together on a sunny day, in a protected wood – where it was safe – on a bank, then they could tell each other lies, they could fabricate histories, they could say something did happen, which did not happen, they could boast. They could talk about what they might have done and could do. They could take each other by the hand and go for a walk, and the planet would tremble and the trees suffer a gentle frisson, and all this only because these sisters passed.

Then they could say as Snow White said to her sister, "Once I let down my golden hair. I didn't compromise, I put on no disguises. I walked out into the world, straight into the street. I was taller than the men, and obviously stronger; but no one stared. I walked into a bar, I sat down, I had a drink.

The people there treated me like a long-time cus-
tomer – in good standing, and the bartender did not
call me 'Sweetheart'. I occupied space and did not
know that I occupied it."

Then Rose Green said, "Sometimes I would like
to fly. Sometimes I would like to lie down under a
tree naked, and have the ants not bite me, and the
sun not be too strong, and no one come along and
think, 'Ah, a naked woman. How nice for me. Now
I can rape her.' I suppose, really, I would like to be
boss of the whole planet and have it solely for me."

And Snow White said, "What about me?"

Or they could say what Alice said. Alice's sister
was reading a book, so Alice muttered, "The cat is
my sister, or if not my sister, then at least my
friend." But the cat protested, "When I torment
birds, are you still on my side? Are you my friend?
You've robbed me of so many pretty playthings.
You're not my sister. The leopard's my sister, and
you wouldn't dare have the leopard as a friend."

## II   SWEET SISTER SOL

i)  BEING LONELY
She says to the moon,
            Are you my sister?
To the frog, Are you my sister?
Even to the lily, so remote, so cool,
      Could you?
            Please?
                  Be my sister!
When the water is still, she could look
for her sister; but the tadpoles laugh:
A drowned woman?
            To be her sister?
She turns to the rock,
            but the rock says nothing,
not even when she carves
            a ledge for herself.

ii)
Against whom are you my sister?

## III   LADY SHY

i)
In the mirror my sister does not smile
at me. She looks anxious and needy. "Sister!"
I call out, "Let's be friends!" She wants
to be friends, but her eyes are like holes,
her voice comes through the glass darkly.

ii)
That night when she came to ask for a cupful of
blood, she did it with charm. "It's not that I'm a
shyster," she explained shyly, "and anyway, between
us there are no debts. It's just – well – it's just –
well," she finished simply, "my need is great." The
next night naturally there was no need for speech.
She extended her cup. I filled it up. Her need, after
all, was very great. And to gain so much virtue for
so little blood . . . The third night, I confess, I felt
some reluctance. But how admit it? The cup was
there. The cup was indubitably and irrefutably
there. And after that, as night succeeded night, habit
made it easier and easier. Am I done for now? Yes,
of course, I am. But before anyone takes me for a
suicide, let the question be asked: What precisely
was the choice I had?

## IV   THE BLACK PIGLET

i)
*Death bade me welcome,*
*But my soul . . .*

ii)
Go to the long beach to meet your death.
There your sister will greet you, take you
by the hand, henceforth lead you. All your life
there was only one other. Your shadow, you say?
No, her shadow,
            not yet made manifest.

## iii) THE BLACK PIGLET

*And this little pig was born in a forest,*
*This little pig was born to run wild.*
*This little one wept and wept –*
*her need was the sorest.*
*And this little one, and that little one,*
*Whoever they were, whatever they did,*
*All of them died.*

The black piglet, having pondered these things, gave up the world, gave up getting and begetting, and took herself off to the scrublands. It's true she had a long pedigree, a good home. But what use was it to be pampered and petted, fattened and fêted, when it all ended with the butcher's knife? What was the point? She would seek out Death, have a word or two with her, make her see reason, somehow make her negotiate. But what could she offer? Only herself. And what could Death offer? She suppressed the answer. There must be a way of dealing with death.

As she nosed among the bushes and turned up stones, she saw a black shadow out of the corner of her eye. A perspicacious piglet, she realised at once that this must be Death. She turned around, but Death swivelled too. "A coward," she thought, "won't talk to me face to face. Or perhaps Death is shy? And this isn't the right time or the right place?" She began wandering through the scrub again. She was aware all the while that Death trailed along. At last she came to a shady tree. Perhaps here they could sit down and talk face to face. The piglet looked around. Death had disappeared. "Where are you, Death?" "I'm here," Death replied. "Where?" "Right here." The piglet shrugged. Even if she couldn't see Death, at least she could talk to her. She had better seize her chance, but before she made a direct request, it might be sensible to find out about Death.

"How are you Death?"

"I'm fine. A little tired."

"Why are you tired?"

"Because."

"Because what?"

"Because I followed you."

"But why did you follow me?"

"I belong to you. I have to follow you."

"But why?"

"Because I'm obedient."

"Are you a dog? A little black dog that follows a piglet?"

"No, I'm a piglet."

The black piglet wasn't stupid. She had begun to understand. "I see," she said. "You are my personal death, and that's why you have the form of a black piglet?"

"Yes."

"But please, Death, I don't want you to follow me about. How would you like it if you were always being shadowed by a black piglet?"

"How would you like it if you were always being dragged by a black piglet?"

"You mean – "

"Yes!"

"Well, let's get a divorce. Let's separate."

But even as she said this the black piglet was sure it wasn't going to work.

"No."

"Why not?"

"Because then we'd die."

The piglet had already guessed at the answer, but she had wanted to force Death to say it.

"All right," the piglet said. "We might as well be friends. Sometimes I'll lead. And sometimes you can be the one to lead the way."

Death was reassured and crept a little closer. The sun beat down. The piglet and her death rested from the glare and fell asleep in the black shade.

## V   I AM ALICE

i)

I am Alice. I am not frightened. I am not vulnerable. The White Rabbit peeps through the looking-glass. He has said he is harmless. He is only a voyeur. He is only a rabbit. In Wonderland he has separated me from both my sisters. But I am Alice. I know I am Alice. I can argue with his creatures. They are his creatures. He can only watch.

The Duchess is not my sister. (Sweet Duchess will you sing me a lullaby?) The Duchess can't sing. She can't mind babies. And her use of pepper does not make her any sort of hot shot. And the Queen of Hearts is not my sister. (Would that make me a Queen, or only a princess?) I have no aspirations (of that sort) yet. And besides, she is absorbed by roses and knaves, and by the king, her husband, though she did ask me to a game of golf.

There's nobody else. Is there? There's the White Rabbit's maid, but I am not the maid. And I'm certainly not the maid's sister. There's the pigeon of course. And the cat, but the cat is neutral. And a snake cannot be a pigeon's sister.

Why has he deprived me of all my sisters? And how does it help? If I were the only girl in the world, and you were only the White Rabbit . . . It is ridiculous. He knows that himself.

I shall run so fast, and race so hard that I shall be a Queen, a fellow Queen with my sisterly Queens, and then it will all come out right.

ii)

One day, in between the pages and slips of the pen, when the cat was asleep, Alice approached her. "We are unobserved," she informed the cat. The cat didn't move, but then again, she didn't vanish either. With due diffidence, Alice addressed the cat, "If both of us were tigers?" The cat smiled. Alice realised that the cat was practising; she was sitting in the sun acquiring stripes. "I am more like a tiger than you are." "Yes." Humbly, Alice conceded the cat's point. Lowering herself gently to the floor, Alice began to stroke the cat. "Sometimes," she murmured, "even if you really were a tiger, sometimes I think we could still be friends." The cat blinked. Emboldened, Alice ventured further, "We must clip your claws." "We must," said the cat, "cut you down to size." Alice retreated. She got to her feet. "Different, but equal?" she propositioned the cat. It was straight from the shoulder, but to this there was no answer. The cat was not a moralist. The cat was not a legalist. The cat was not concerned with egalitarian attitudes. Carroll the Creator had got it right: the cat lived from moment to moment.

iii)
Tumbling down the black hole,
      awaiting yet another re-birth,
even in this vacant and unborn state,
      I am not passive,
        I am voracious.
There but for the grace . . .
      Millie can't spell.
       Tillie is tedious.

## VI   VERITY AND CHARITY

i)
The sisters are calling.
Weary wanderers? Soiled seafarers?
No, not them. Only each other.

ii)
These sisters:
    Let them be just and generous.
    Let them be dispassionate.
    Let them speak
        (without any pleasure)
            of the deaths of men.

iii)
Then Verity and Charity set off together
into the sunset.

## VII  O MEMORY

*O Memory, Mother of Muses,*
*Let your daughters grow up.*
*Don't make them all shut up.*
*Let them re-remember*
*The babble of bygone days.*

Three Graces, nine Muses, seven starlets, ten magpies – were the magpies sisters? Yes, why not? – and one gosling were all dumped together in a leaky boat and sent off to meet their Fate. How do you do? How do you do? How do you do? No cock-a-doodle-doo now. No further adieu. The tide slurped and slurred their sounds. O joy. O new found land. O pie in the sky! O blackbirds freewheeling freely about the heads of the Fates, who also being sisters, were very like them. And nary a King in sight, not a whisker of a King, in their new found state. The sea, maudlin and malleable now, tossed them back their new found sounds like pebbles on the shore, and the pebbles grew and grew and no pebble ever stayed the same. "This is my pebble," said the little gosling, seizing one that was still wet, and even as it touched her golden beak it turned into a golden egg. "Ma!" cried the gosling, "Ma!" and snuggled up close and the magpies laughed and feathered her bed. Later, the gosling began to grow black and white feathers, a Canada goose, a black and white magpie, an un-sistered bird? No, no such thing. A transitional phoenix, for an egg is a seed, is a tree, is a tufted bird, rising high in the sky with other feathered creatures whirling round and round and round her head. And later the sea was chopped and cut up, the sacrificial goose was cooked and carved, and the pebbles became words on a page, on which the sun shone and shone and shone forever like a golden egg.

What language did these sisters speak? I do not know. Very soon and very quickly my vision ended.

## VIII   SWEET SISTER SUCCESS

*Be good, Sweet Maid.*
> *And let those who can't –*
>> *grumble!*

i)
Step-sisters compete, but then so do sisters. They compete even harder. But this girl, Cindy, or Cinders if you like, had two brothers. These two brothers were astute and ambitious. "Cindy," they said, "since the shoe fits, you've got to wear it. We seek preferment. The Prince would make an excellent patron." And Cindy, also being astute and ambitious, did what was asked, to the end that the three of them lived happily ever after.

ii)  RAP RAP
I will climb to the top of the ladder. I will do push-ups. I will not hiccup. And when I reach the top of the tower, then I'll let down my golden hair, and all my sisters can join me. They will say, "Ah yes. Yes! She was the first! She is a hero!" But I fear in the end they'll learn to accuse me. "A witch," they will say. "She has set up claims. She has insisted on rights. She has said once or twice she would like her privacy."

iii)

Little Red, Ruffianly Red, sweet, smiling and impoverished Red, with the wolf knocking constantly at her door, and not a man in sight, just an endless, mirroring series of mothers and grandmothers, who were once little girls, who each in her way has had to deal with wolves, who are good for advice, also home-cooked meals, but who, in their myriad manifestations and multiple histories, have left the problem of the wolf unresolved, so that wolves are everywhere and Wolf is a fact, Red must now decide what to do for herself.

"Brother Wolf," says Red softly. (There's a chain on the door and at first she is tentative.) "*Brother* Wolf." There's appeal in her voice. But Wolf is not a good Franciscan. The wolf is a literalist, and Sister R is good red meat. That is what she is in herself. The rest is a matter of text and context.

The door is still barred. But there's a wolf at the door. And the cupboard is bare. Sister Red has to try again. She touches her fangs and concentrates hard. "*Fellow* Wolf!"

The Wolf growls, "If you want to join the gang, you've got to sign up in flesh and blood. Who else have you got hiding in there?"

"Only grandam."

"Fetch her out."

But Little R can't quite bring herself to do that. She thinks again. "Hey, Wolf!" she calls out.

"You're not a wolf. You're really a man."

"So what?" says the wolf.

"Marry me!"

"What good will that do?"

"Then we'll have kids and farm the land."

"Have you got any land?"

"Yes," replies Red. "But I've got to clear it of wolves first. I need your help."

So Ruffianly Red and Erstwhile Wolf Man make an alliance. They buy shotguns. They work hard and clear the land.

## IX   GORGONS! AHOY!

i) THE DRIFT
Meanwhile, two sisters in a boat
        slowly approach a darkening shore:
"There's an island called 'Death'," says
                                one sister.
"No, Death is a continent, an entire planet.
        Nor are we off it," replies the other.
The boat drifts; the light dims.
                        Or only their eyesight?
"Seize the day? By the end of its tail?"
                        suggests the first.
"Look, there's the smallest slit of daylight."
"The beauty of the sunset? Shall we enjoy it?"
                        murmurs the second.
In the boat there are provisions,
                        even some wine.
She rummages slowly. It's the drift
        of the boat – slows things down.
"Strange, is it not," enquires one sister,
        "that we're both in this boat?"
The other one shrugs,
                "A matter of metaphor."
She smiles at her sister and takes her hand.
                They sit in the darkness.
The boat drifts.
                They stop holding hands.

ii)

But who were these sisters? Hags, crones, two old women drifting off without ever having said a proper Good-Bye? The two eldest Gorgons, absconding, as it were, and leaving poor Medusa to fend for herself? But suppose young Medusa won't stand for it? Suppose she dives straight into the water and swims off strongly in pursuit of them? Suppose she catches up. Suppose she says, "Hey, you can't go off. I still need your help." What can these sisters do for her? What can they offer? Only their boat, which Medusa, who by now is tired and breathless, no doubt accepts.

## X   AND NOW, ALL TOGETHER!

Alice stepped forward, "There is a problem. The problem is I haven't any sisters. I am unique."

Then Lady Shy: "There is certainly a problem. The problem is that the sisters aren't sisterly."

And Medusa said, "No! The problem is they aren't motherly! The Sisterhood of Man – it sounds grotesque! Look how the snakes writhe in my hair! Mirror-eyed sisters! Look at what you've done to me!"

Then Rap Rap spoke, dispassionately, divesting herself of vested interests, "The problem is that not all the sisters are equally clever, or nice, or amusing."

Sister Verity nodded, "That is very true."

And Sister Charity added: "But how does it matter? And what does it mean?"

Meanwhile, The Black Piglet: "Death is my sister, and yours and yours . . ." But this was so obvious, nobody knew what to do with it.

The thin air parted; the cat spoke, "I'm more like a tiger than you are. You're more like a monkey than I am." Everyone heard; they tried not to notice.

And Little R roared, "I'm covered in red. Red blood of Grandam. Red blood of Wolf. Even my own red blood, probably. I don't like being Red. You've stereotyped me!"

"Well, but my dear –" Cinders simply – "The problem is we haven't much power of our own, you know. We have to be someone, do something."

"The problem is we've all been typecast!"

"The problem is God!"

"No, no, the problem is man!"

"The problem is us. For original sin/ Look within."

"The problem is mother!"

"The problem is words!"

"We need new words!"

"Yes! Words that are strong and slippery as eels."

"Yes! Words that mate and proliferate."

"Yes! Words like pebbles, thick on the beach."

"Yes! But not like pebbles. Words like sweets."

"Words like sweets and seeds and pebbles."

"No, not 'like', words that 'are'."

"And aren't as well!"

"Words that swim through the seas like fish."

"And flip into the air!"

"To make a dainty dish!"

"Words we can eat!"

"Words we can drink!"

"That won't poison us."

"That will poison our enemies?"

"'Enemy' is a word."

"Why so it is."

"We will build a great edifice out of malleable words."

"We will cultivate a culture."

"We will grow a common language."

"The Sisterhood of Women will mean something!"

"We will bring our own words."

"We will own our own words."

"We will be able to make anything!"

"Anything we make will be something!"

And that is how it came about that the sisters began building Babel again, for no real reason, except perhaps an urgent need.

# II

## THE LIFE AND DEATH OF
## THE BLACK PIGLET:
## SOLLY'S ACCOUNT *(duly amended)*

*These fragments they shored, shaped,
sharpened, shaved, and in deep humility handed
over to be worked by Time.*

### 1

It would not be a tower and it would not be destroyed. It would be different, but nevertheless, some planning was required, some minimal organisation and apportioning of responsibilities; and so it was agreed that Rap Rap would be the architect, Little Red would find a suitable site, and Cinders would deal with the planning permission. The function of the others was to provide raw materials. In this way, though the tower, the reef, the mainframe complex or the subspace satellite, whatever it was that it would ultimately be and perhaps go on being, was as yet unbuilt, the sisters achieved a connectedness, even the loners, who could not help taking an interest.

Little Red, though unwilling to make use of her own assets, viz. the lands she had cleared with the aid of Wolf Man – he would object, her children would object – did some research, set aside time and travelled some distance in order to find Sister Solitude. She found the latter on top of a sand dune. Behind her stretched miles of sand, before her stretched miles of ocean. She was sole mistress of all she surveyed and nobody questioned her rectitude.

Little Red approached, settled her own firm flesh about her bones, and glanced at the scarecrow perched overhead. She cleared her throat, and came to the point: "The sisters are building Babel again. Could you please give us some land?" *[I had been prepared for coldness, coyness, a bit of rudeness even, but for what happened next I wasn't prepared. Red]* Sister Solitude opened her mouth, and words came out *[odd words, in clumsy garlands].*

"The fish has no sisters. The bird has no sisters. Many mammals too . . . but let's talk about fish, mud fish, angel fish, the kind of fish who change their sex should the need arise. These fish, they perceive members of their own species, male and female, and unhatched eggs . . . And the squirming young? Are they perceived? And if perceived, are they eaten?

I, Sister Solitude, perched on a sand dune, am attempting to understand the bounds of bonds. I have left my own community. I am trying to think. This may be heresy. But in the interests of truth . . .

The Eater and the Eaten, the I and the Other. Surely Eaters are members of my own community? And yet, what I ingest is surely an inalienable part of me?"

At this point Sister Solitude *[looked arch and]* smiled dazzlingly. *[And me, Red, not knowing what to do, nodded intelligently. This worked, because Sister Solitude went on happily.]*

"Well then, Eaters and Eaten form a partnership. They are component parts of a singular species. And between the Eaters there has to be rivalry. Then which is who and woe is me."

This time Sister Solitude looked *[both arch and]* expectant. And this time Little Red who was beginning to get the hang of it, ventured a reply: "If I look in the mirror and like myself, then I must like all the others who look and like, walk and talk, just like myself. Thus we may live in amity."

Sister Solitude nodded and responded almost instantly: "Then Cain and his brother must love each other, and Cain and his sister must hate one another."

But this fazed Little Red, and she hesitated. What about Wolf Man? What about her sons? She tried hard to think of something; but her hesitation was just what was wanted, because Sister S concluded in triumph:

"There is no other predator, she sings,
      she sings so sweetly,
            only you, only you . . .

Yes, you may have my land, and my poems. Let that be my legacy."

With that Sister S strode off in search of yet another desert; and Little Red gathered up the words and the title deeds.

Cinders, meanwhile, was pleading with her husband, and with her brothers. "Please," she said. She said it so nicely. "I have a whim, a fancy, a fantasy."

"What is it, my dear," they all of them said, "and what may we do to gratify it?"

"I want to set up a little project, a centre for women who wish to learn and work very hard."

"Do you mean a university?"

"Well, no."

"Do you mean a factory?"

"No, no."

"Do you mean a nunnery?"

And when Cinders began to shake her head, they asked her what she did mean.

She replied, "What I had in mind was a palace in the air and under the sea, a structure that was both real and impossible, in short, a functioning monument."

The prince and her brothers were taken aback. Recovering himself, the prince murmured, "I'll have one built – right away if you like."

Cinders protested, "It's the building of the building that *is* the project."

The three men were genuinely baffled; but the

prince was uxorious, and her brothers were brotherly. "Very well, my dear, what is it exactly you want from us?"

"Your permission to build, and access to materials in order to recycle or to invent."

"You have it," said the prince. Her brothers nodded. "But you haven't asked for much."

Cinders said nothing. She had begun to realise just how much she had asked.

*[Solly, this is hardly objective. You make us sound like plotters and planners, who had no real idea of what they were about. Once Babel was begun, do you seriously expect anyone to believe that the men sat back and allowed the towers and minarets, the houses and fields, to rise and flourish without their help? You don't do me justice. Someone had to deal with the problem of men. In sisterhood, Cindy]*

Little Red, Cinders and Rap Rap were efficient women. Cinders informed Red. Red informed Rap Rap. Rap Rap informed Cinders. Permissions to plan and to re-invent were received in writing. The title deeds were safely stowed. And Little Red and Cinders agreed to meet Rap Rap at Solitude's sand dune in three days' time. When they got there, they found that the sand dune had disappeared. In its place was a mountain, which appeared to consist entirely of garbage. It was not that the individual items were garbage, so much as the state that they were in. Rap Rap had climbed halfway up.

"Your husband sent at least half this mountain," she informed Cinders.

"And the other half?" enquired Cinders.

"Sent by well wishers and do gooders, by sisterly sympathisers," muttered Rap Rap. She frowned. "We must inform the sisters," she continued crisply, "that a major part of Project Babel exists in cyber-space, and that the materials they send must belong to that realm as well."

"What on earth do you mean?" asked Little Red.

"Well, it's obvious, isn't it?" replied Rap Rap. "Essentially we're engaged in constructing a dic-tionary. For a great many things the words will do, we don't have to have the things as well."

Cinders intervened. "We are not engaged in con-structing a dictionary or even an encyclopaedia. We are constructing the universe itself!"

"Can't we just construct civilisation, and leave the universe to its own devices?" Little Red inter-jected. The sun was beating down. The mountain was mountainous, and surely a little moderation was just common sense?

"I thought you were constructing the sister-hood?" This came from Solitude, who, unable to find an empty desert, had returned to the sand dune.

"That too," Cinders muttered.

The four sisters felt disheartened. Little Red sighed, "Whatever it is we're actually building, it's going to take a very long time."

Right then and there, at the foot of the mountain, they held another meeting, and with celerity and unanimity and even some amity, they faxed The Black Piglet and invited her to sort through the mountain that towered overhead: "RAW MATERIALS READY. MOUNTAINOUS TASK. ONLY YOU CAN DO IT. PLEASE ACCEPT."

Faxes flew back and forth. In the end The Black Piglet accepted their invitation provided she was given a great deal of help. However, when The Piglet arrived, she found that Cinders had had to return to the palace, Little Red to her homestead and Rap Rap to her tower. Only Solitude remained, who for reasons of her own, offered to help The Black Piglet.

*[Babel wasn't built by a favoured few cogitating on the cosmos. Describe their glory by all means. But it takes Plain Speakers and Sooth Sayers – me, for example – to tell the truth. Who was The Black Piglet? What is her history? Her date of birth? What was her character? Had she any flaws? What you have failed to tell is the truth about Babel. In all sincerity, Verity]*

Solitude and The Piglet gazed at the sand, gazed at the sea, gazed at the mountain, and at each other.

"Right," said The Piglet, "we sort everything here into two categories."

"To do with Women and Not-women?" asked Solitude helpfully.

"That's passé," replied The Piglet. "And it's parochial. And in any case, I'm not a woman, and you're not a not-woman, so it would all be a mess."

*[Aha! Got you! Did The Black Piglet really say that? If she wasn't a woman, who was she then? Solly, I strongly suspect that when you aren't really sure, you just invent. Verity]*

"Let's put everything to do with words into a pile over here."

"That's only one category."

"One is always two," The Black Piglet informed Solitude. "There's words, you see; and then, of course, there's everything else."

With her customary single-mindedness The Black Piglet began rooting through the mountainside. Solitude helped. Computers, televisions, printers, telephones and old newspapers she had no problems with – they had to do with words. But then she had to pause. "What do I do with three odd socks? After all, these things are words too."

"Use your discretion," snapped The Piglet. She considered Solitude's work habits lax.

Solitude made another little pile, which in her own mind, she labelled "Contributions"; but after a few minutes she stopped again. "What about this brick?" she called out to The Piglet.

"Has it a message?"

"I don't know," replied Solitude.

"Well, chuck it here," replied The Piglet; and to

save interruptions, she informed Solitude that all bricks either could or did or even might or should, carry a message, and that therefore all bricks should be saved.

And so, they laboured and slept, and slept and ate, and laboured and ate, and slept and laboured. At the end of forty days The Piglet pronounced their task completed.

Solitude looked about her. Instead of one large mountain, they now had one fairly large mountain ("To Do with Words"), one smallish mountain ("Not To Do with Words") and a little mound ("Contributions"). Was this progress? Solitude didn't, however, contradict The Piglet; and The Piglet, without any hesitation, faxed Rap Rap: "PHASE 1 FINISHED. SEND TECHNICIANS. SEND CINDERS. COME YOURSELF."

*[Methodology! Oh well, at least the faxes constitute some sort of documents. Rap Rap]*

That night as The Piglet and Solitude waited for Rap Rap or Cinders or Little Red to put in an appearance or at least fax back, they found that for once they had time on their hands and they began to chat. Solitude was quibbling with The Piglet, "No, crone is *not* derived from Chronos. Don't tell lies."

"Why not?"

*[Sol to Rap Rap. This is first hand evidence. Don't you understand? We had time on our hands, in our*

39

*ears, our nose, our throat. We were drowning in time. We had to say something, do something, to get ourselves out of the morass.]*

"Well, at least tell lies that are connected to other lies, so that then we have a beautiful and complicated system."

"All right." The Piglet obliged. *"Once upon a time there was a beautiful old woman called Crone Kronos."*

"No!"

"Why not?"

"Crone Kronos! Widow Dido. It doesn't sound right."

"Sounds fine to me." The Piglet continued. *"Poor Crone Kronos was, as it happened, barren: no children, no future, no variations on a theme. She wanted a change. And so – "*

"Parthenogenesis, I suppose?"

"No. *She chopped herself into infinitesimal pieces with which she imbued the universe, and that is why the universe is imbued with time."*

"And that is why, I suppose, everything in the universe changes so? Dame Kronos' stamp? No, I know a better story than that about her. And this one is factual." Solitude smiled authoritatively. "You see, I actually knew the old lady."

"Which old lady?"

"Crone Kronos."

"But I thought you said that couldn't be her

name?" Piglet protested.

"Don't quibble. Yes, *I worked for her as her servant girl for a whole year. She was tough, that lady. I had to fetch wood for the fire. Boil water for her bath. Fill the bucket with water. Make a gruel for breakfast, shake out her bedclothes, remake the bed, sweep the floor clean, go get the groceries, make a snack for lunch, dig the potato patch, see if the hens were laying, go to the well for water, trim the lamps for nightfall, fix a meal for dinner, clear and clean the dishes, and then tell Crone Kronos a bedtime story even though she never slept.*"

"And then what happened?"

"*At the end of the year I asked for my reward. I thought that that was what was supposed to happen. But Crone Kronos just laughed at me. I asked her why and she replied, 'You stupid girl, all that happens now is that you have permission to work for another year.' So then I said, 'But when do I escape?' And she said, 'When I'm bored with you, dear.'*"

"How did you escape?" Piglet asked.

"I didn't escape. I made a deal. I put it to her, 'Look, I'm not much good at making beds or washing dishes, can I just tell you stories instead?' She agreed. So now I spend my time making stories for her."

"That's funny. I do the same. But what does Crone Kronos do with your stories?"

"Oh, she re-invents them. But then that's pretty

much what she does with everything. This story I'm telling you, it's one of the ones I made up for her. Do you like it?"

"Well . . ."

"Change it then. Pretty soon Crone Kronos will come along and change it again. That's how it goes."

With that Solitude rose and went off to sit on a nearby rock; but when a penitent Black Piglet approached Solitude and asked why she was sulking, Solitude said she wasn't sulking, she was making up yet another story. Why? To placate Crone Kronos, of course. The Black Piglet did her best to look respectful. It was lonely in the desert, and on the whole she quite liked Solitude.

*[Ah yes, you are "penitent", you are "cross", you are polite; at worst you are petulant. But was the sun hot? Did you scent Crone Kronos on the desert breeze? What was it you really felt? Are you so alienated from everyday life and ordinary women? In exasperation, Lady Shy]*

The following morning they heard the sound of a boat being dragged along the beach. They ran down to look. When they got there, they found two elderly women, who introduced themselves as Verity and Charity. They had come to help. Rap Rap, they said, would be arriving shortly, and, indeed, when they got back to their camp, they found Rap Rap there examining the small mountain, the smaller mountain and the hill. She had expected order, a plan, and an outline; but she had found the sand littered with stories instead. Briskly and silently she swept these into the mountain. The others watched. The problem was obvious.

"The sorting you have done has been remarkably efficient. But now we need a further classification, a clarification, of the materials at hand." Rap Rap spoke quietly. She sounded tired.

For a moment they all had a vision of miles and miles of tiny mountains. Something simpler was needed.

"Well, there's True and False," offered Verity.

"Or Cruel and Kind," offered Charity.

The Black Piglet sighed. "The fact is there are no simple subdivisions, no distinct definitions. There are only degrees."

"Degrees of what?" asked Rap Rap impatiently.

"Of things more or less infected by Time."

Solitude opened her eyes wide. "Infected!" she breathed.

"Well, 'imbued' then," amended The Piglet.

*[Now, Solly, what does it mean to say "infected by" time? Do you mean "subjected to"? And why "more or less"? Rap Rap.*

*Little Red to Rap Rap: I think "more or less" means that a stone is less subject to time than you are. Let's not get into metaphysics. All this is going to be difficult enough.]*

43

"But why did you sort things into 'Words' and 'Not-words' ?" Rap Rap demanded.

"Because Words are tools."

"Tools for what?"

"For grappling with time!" Solitude put in.

"Well, but what about that pile of equipment over there?" Rap Rap asked, pointing to the mountain.

"Those are tools for the tools," The Piglet replied. "There's enough there for a mainframe computer, any number of PCs, a television tower and a tele-communications satellite. It's beginning to rust. Will the technicians arrive in time?"

"Yes, indeed. There they are," Rap Rap replied indicating dozens of vehicles scurrying across the sand like giant insects. But Rap Rap was still worried. "What about that mountain?" she asked pointing to the pile of "Contributions".

"Oh, I'll deal with that," volunteered Charity. "The workers will need to be sustained and fed."

"And what will you do?" asked Rap Rap, turning to The Black Piglet, not daring to impose on The Piglet further, but hoping that she and Solitude would stay to help.

"We will sit on top of that dune over there," replied The Piglet solemnly, "and tell each other tales of the workings of time."

"But how does that help?'

"It is crucial to the project." The Black Piglet stated this with so much conviction that the others did not dare challenge her directly.

*[Well, that's true enough. Red]*

"What will you do with these tales?" ventured Rap Rap.

"We will inscribe them upon those bricks, which in time will house all our projects," replied Solitude.

"And I will be in charge of all the bricks on which the tales of time are inscribed and I will pronounce on the truth in them," put in Verity. She had fluffed out her voluminous garments in such a way that she managed to look larger than life.

"How will you do that?" inquired The Piglet.

"I have my methods," replied Verity. "Some are democratic. Some are élitist. And some are limited. There's a truth for every occasion, and an occasion for every truth. Leave it to me. I know best."

*[Verity, what has already been inscribed and is even now being amended, extended and even erased, must be done by consensus. Please remember that. Rap Rap, Cinders and Little Red.]*

Verity, Solitude and The Black Piglet set off towards the sand dune, Charity began sorting through "Contributions" and Rap Rap stood where she was and watched the technicians put this into that. She felt muddled. What was whatever they were building going to look like? In order to make herself feel better, she called out after The Piglet. "Will the bricks on which you inscribe the tales of time be inscribed on only one side and will that side be on the inside?"

"On all six sides," shouted back Verity. "Truth is open to the broad daylight, is dark and hidden, is sometimes half seen and half understood, is sometimes obliterated, The truth is truth is multi-faceted!" It was obvious that Verity was enjoying herself hugely. *[Well, why shouldn't I enjoy myself? Oh, and by the way, Solly, I don't think you gave my sister, Charity, a large enough part. Verity]* Rap Rap turned away. She faxed Cinders: "NEED HELP".

*[You say I felt "muddled?" Can you blame me? At least concede that with whatever you gave me, I did my best. Rap Rap]*

## 2

*One evening Crone Kronos was taking the air with a group of her companions. They walked through meadows, forests, hills, and wherever they walked, it soon became apparent that the little flowers, including the buttercups and daisies, and even the coarse blades of grass, were swooning and languishing at Crone Kronos' feet.*

*"Nature adores you," murmured one of her companions.*

*"Don't be silly," snapped Crone Kronos, "what you observe is not adoration, it's primitive greed."*

*"But what do they want from you?"*

*"Mercy," replied Dame Kronos. "They think I'm Dame Plenitude and want to feed off my body."*

*"And aren't you?" enquired the young woman gazing with adoration at Dame Mercy.*

"How's that?" asked Solitude when she'd finished. "Keep or scrap?"

"Keep, I think," replied The Black Piglet realising that that was the end of the story, "but this technique of yours – rich in questions and sparing in answers – don't overdo it."

Verity had listened with attention. Now she began to mumble. "Buttercups and daisies – yes. They bloom at the same time. The nature of the old lady's companions not specified – therefore hard to prove misinformation. I'll let that pass. But what I do object to, Solly" – Verity had a way of taking liberties – "is the utter confusion in terminology. Who are Dame Plenitude and Dame Mercy? If so, are they the same as Dame Kronos? If not, are they the same as each other? And in any case what is the relationship of Dame Kronos to Crone Kronos? Oh, and how precisely is 'Kronos' spelt? And what has any of this to do with sheep? I'll tell you a better story, and what's more I had a part to play in it."

*Once upon a time Crone Kronos had a favourite, a sculptor called Lovely. Now Lovely was clever, disciplined and sensible. She tended to Dame Kronos, brought her flowers, sat at her feet, and made it clear every day and in every way that she was Kronos' disciple. And so, when Lovely stood before a great rock face and began to carve her images upon it, Crone Kronos stood there behind her and (it is said) helped her a great deal. Indeed, even after Lovely died (it is said) Crone Kronos can be seen from time to time at work on the rock face, altering and improving, so that Lovely's fame might continue to live. And that is why it is said of Lovely that she was the greatest sculptor who ever lived.*

*[Solly! That's a really good story. And whatever you say, I see that you thought well enough of it to take the trouble to record it. And I don't take liberties. Even your account, biased though it is, makes it perfectly clear that I have always tried to be helpful and friendly. Sincerely yours, Verity]*

"Wasn't she the greatest sculptor who ever lived?"

asked The Black Piglet.

Solitude didn't know that The Piglet was capable of being disingenuous. "That isn't the point," she interposed seriously. "The point is she was Crone Kronos' friend, isn't that right?" Verity nodded, but The Black Piglet had a further query, "What was the part you played in it?"

Verity looked coy. *[I am not coy! Careless, perhaps. Casual sometimes. Influenced at times by my sister, Charity. But coy – never! Still in sisterhood, Verity. And by the way, Solly, I've been thinking about it. That story you told about having been a servant to Crone Kronos – that's outright fabrication! You! A landowner! I don't believe you've done a stroke of work in your entire life!]* "Well, the Test of Time – for any work of art – is, after all, the Test of Truth, don't you see? Anyone who's a friend of Crone Kronos gains a great deal of credence with me."

The Black Piglet sighed. *"I knew a fellow once – a woodcutter – who had also wanted to be a great artist. He had come to the forest to think about it. He went a little mad. He said he was going to tattoo the body of Crone Kronos, copy the rhythms, feel the heartbeat – the circulation of blood. He was, he said, going to murder the old lady and dissect her body. But each time he did it, Crone Kronos slipped away and it was someone else's carcass he had to deal with."*

"What a horrible man!"

"Yes. Well." The Black Piglet shrugged.

"I myself am a vegetarian," Solitude put in.

*[Blood and guts at last. But you're not really willing to deal with reality. I accuse you, the pair of you, of cowardice! Lady Shy]*

*Rap Rap to Lady Shy: Look, we've allowed you a say in what's kept and what's cut, but your concern with blood is downright obsessive. What is the point?*

*Lady Shy to Rap Rap: Death. Life and Death – that is the point. Can't you see that the chronicle of The Piglet is all about Death, but that Sister Solicitude fails to admit it.*

*Sol to Shy: Are you Lady Death?*

*Rap Rap: Sisters, stop it! You are reducing our work to the level of graffiti.*

*Little Red: Yes, graffiti scrawled on the walls of Babel, etched on brain cells, concealed in megabytes. What's wrong with graffiti?]*

After a light repast the three women returned to their task. Verity cleared her throat. She seemed embarrassed. With a diffidence that suited her not at all, she asked hesitantly. "Excuse me, I don't in the least doubt the value of what we are doing, but just between the three of us I have a question. Would you mind telling me: Does Crone Kronos really exist? I mean is she a goddess or a person? Have you actually met her? What does she look like?"

*[There, you see, I asked the right question. What would the two of you have done without me? And I can answer that. You'd have done nothing! As for diffidence, hesitation – I can be as humble as anyone else! Verity]*

The Black Piglet and Solitude exchanged glances. Then Solitude said, "Crone Kronos is a goddess. She has many shapes. Now could we get on with our work please?"

But Verity had noted the exchange of looks. She didn't like being patronised. It made her obstinate. *[Not obstinate, persistent. Verity]* "Yes, but what exactly are we doing? Why are we telling the Tales of Crone Kronos?"

"In order to achieve a vision of our undertaking," The Black Piglet muttered.

"Do you mean of Babel?" Verity whispered. She was impressed. She ventured one more question. "Will Babel be the Temple of Kronos?"

The Black Piglet stared. "Yes," she said. "Yes, that is exactly what it will be . . ."

For a while no one said anything. Then Verity started up again, "And words, words are the bricks of the temple, isn't that right?"

But here The Black Piglet shook her head. "Not exactly. Words are the vessels of Crone Kronos. We unseal the vessels in the Tales we tell."

"And then?"

"And then Crone Kronos and her retinue march

past our eyes."

"Are we making a film? A kind of documentary?" asked Verity.

"Well, yes, but the medium isn't film. The medium is memory."

"Who is Memory?"

"Memory, mother of the Muses. Crone Kronos' step-sister."

"Yes, well, I see. The bricks then? I suppose they're megabytes?"

Before anyone could reply a voice interrupted, "It's my turn now. I'll tell you a story about a caterpillar." It was the cat, hovering in mid-air, able to be here, there, or partially anywhere. *[Solly! I don't ask for much, but henceforth, would you kindly spell my name with a capital C. Like anything and everything I am unique. Capital C]*

"A caterpillar!"

"Yes." The cat was adamant. "If words can be bricks and/or vessels, then I don't see why Crone Kronos can't be a sculptor and a caterpillar."

"But how do you know about the sculptor and the caterpillar and all the rest?" Verity asked.

"Crone Kronos is a crony of mine," the cat replied. "In one of her incarnations Crone Kronos was a caterpillar. She belonged to a unique species."

"Aren't all species unique?" interposed Solitude. "I mean isn't that what makes them unique? They are themselves and not something else."

"Yes, well, Caterpillar Kronos was unique," the cat hurried on.

"Was there only one of her?" Verity interrupted. She wanted everyone to know that she liked to be precise.

"Yes. Crone Kronos was a unique caterpillar. After all, isn't everything unique – if you think about it – in so far as it isn't something else? If you interrupt again, I won't go on."

*"Anyhow, Caterpillar Kronos exuded a valuable substance from the tip of its nose, which women and men valued highly. It was said that a tiny amount of this substance would confer eternal youth upon the beneficiary. A mighty Emperor offered a mighty reward for the capture of this unique caterpillar, and eventually – the incentive was so great – the caterpillar was captured and brought to the Emperor. He was pleased. He himself had a tiny amount of this strange substance (bequeathed to him by his great-grandfather), but there wasn't enough there to be a suitable dose for an Emperor. Now that he had the Caterpillar itself, he would be able to collect more. The Emperor and his Councillors watched Caterpillar K. It looked sleepy. Indeed, it looked as though it was ready to begin a cocoon. Mustn't let it do that, they all realised. When it emerged from its cocoon, it wouldn't be itself, well not quite itself, anyhow it wouldn't be Caterpillar K. How prevent it? The Emperor fed Cat. K the tiny bit of the substance*

*inherited from his great-grandfather – enough for a
caterpillar, though not for an emperor. Cat. K ate it
and gave up on the cocoon. Eventually it secreted a
tiny bit of the substance, which eventually the great
grandson of the Emperor inherited."*

The c̲at finished and blinked slowly to indicate
this.

*[Pretty, pretty, pretty. The caterpillar does not
decay and does not rot. And works of art live forever
or fade away – without a smell. They were seeking
immortality in these cultured games they played so
well. And as for Crone Kronos, let us split Crone
Kronos, but let us not admit she has a third sister,
Death. Lady Shy]*

"It would appear that your story has a moral,"
ventured The Black Piglet. "Are you suggesting that
Crone Kronos is a true cannibal and self-regenerating,
that is, she derives her nourishment from within
herself?"

"Yes, like me. I derive my entertainment from
within myself."

Verity who had been imagining millions of
Caterpillar Ks, each highly individual and highly
unique, and separated from one another by instants
of time, decided she would now change the subject.
She formulated a question to which she thought she
might want the answer. "If Memory is the sister of
Crone Kronos, then how do the old crone and her
nieces get on?"

"Indifferently," the <u>c</u>at replied. "But the nieces are dependent on their rich relation."

Solitude had to think before she understood Verity's question, but when she did, she nodded approval. *[You see!!! Vindicated out of your own mouth! Verity]* She looked at The Piglet. The latter was watching a procession of women shuffling through the sand towards their dune. Perhaps she hadn't heard? But then she spoke.

"*One day Crone Kronos and Madam Memory were going for a walk. Crone Kronos kept racing ahead and Madam Mem kept lagging behind. At first Crone Kronos attempted a sisterly concern; she rallied Mem, 'Come, my dear, you are much younger than I am. You should be racing, while I fall behind.' Mem just gave her a sour look: her ankles had swollen, her temples were throbbing, and she wasn't in the mood for jocular rallying. Then Crone Kronos tried another tack. 'What you need is a handicap. You go ahead. And when you've gone a long way ahead, then I'll follow.' Mem didn't know whether K was being nasty or just stupid, since she knew perfectly well that she, Mem, only had eyes at the back of her head. She pointed this out, sat down on the grass and gave up the race. Crone Kronos sat down beside her. She patted her shoulder. 'Tell you what, dear, summon your daughters.' Mem did so. Eventually her daughters appeared in ones and twos and threes. 'Now then children,' Crone Kronos commanded her nieces, 'help your mother.' Help? The*

55

*nine daughters looked at one another. Help? Hither-to they had considered their existence ornamental. Besides, mothers were supposed to help their daugh-ters, not the other way about. But they were afraid of their aunt. 'How?' they asked. 'Lead her,' replied Crone Kronos, who was getting tired of the lot of them, and went on ahead. Ever since then the nine daughters of Mother Mem have taken to leading her – sometimes this way and sometimes that, and some-times they just sit by the roadside."*

"But what does your story mean?" demanded Solitude.

"It means," said The Black Piglet, "that Mem is imperfect. It means that sometimes Mem can go – with the aid of her daughters – where Crone K has not yet led. It means that Mem and her daughters can sit down sometimes. They do it when they're tired or disheartened or just feel like it. As the nieces of Crone Kronos that is their privilege."

*[Oh for Crone's sake! Art! Myth! Mystery! Metaphor! It's difficult to know whether to get angry with you, or dismiss you both as a pair of liars! What about life? What about death? And if you don't care, then what about the lived lives, the hopes and fears, the needs and aspirations, of ordinary people like myself? Babel was built with our blood, you know, not with your breath. Lady Shy]*

"And what about Crone Kronos?" asked Verity, not to be outdone in the business of asking intelligent questions. "Where and when does she sit down?"

"On the right hand of God," Piglet found herself thinking. *[How do you know what The Piglet was thinking? It seems to me, Solly, you're doing a great deal of interpreting. You should stick to what The Piglet actually said. Verity]* But all she said was, "That is an extremely difficult question. Crone Kronos is not subject to Crone Kronos. Who she is when she isn't no one can answer."

Verity was delighted. She tried another question. "Has anyone succeeded in charming the old Crone?"

"She picks her favourites from among her nieces," the cat muttered.

"No, not her blood relatives. Anyone else?" Verity was persistent. It was evident she was thinking of herself as a candidate. *[And how do you know, how can you possibly know, just what I was thinking? You exceed yourself! Verity]*

"Well, there's me," the cat offered.

"No, not you. I mean a person."

Offended, the cat vanished. No one said anything. The army of women was close at hand.

Rap Rap, Little Red and Cinders stepped forward. They did not salute. No one was quite sure of The Black Piglet's rank. Rap Rap said, "The technicians are ready. Have you completed The Tales of Crone Kronos?"

The Black Piglet smiled pleasantly. She shook her head. "The Tales of Crone Kronos are a process. To ask whether they're finished doesn't make sense."

Little Red and Cinders looked at one another. It was obvious that The Piglet was going to be difficult. Fortunately, Verity intervened with her customary brashness. *[Solly, be fair! Over and over it was I who managed to save the situation. Verity]*

"You're just in time," she shouted exuberantly. "We've finally understood the nature of the problem. We know what we're doing!"

"And what is that?" enquired Rap Rap.

"We're building the Temple of Crone Kronos!" she replied in triumph.

The sisters looked at Solitude and The Black Piglet, waiting to hear what they would say, but they were gazing into the far distance. Finally, Little Red asked a direct question. "Is that right? Are we building the Temple of Crone Kronos?"

"Yes," replied The Piglet. "Yes. That is right." But she sounded distracted.

There was a murmuring among the sisters. Who was Crone Kronos? How did The Piglet know what to do? And had the decision been arrived at collectively? But when Verity began bossing them around, they did what was asked without further troubling themselves.

"The remains of the mountain are to be used as rubble," she informed the sisters.

"Right," they said, and began carting it away.

"These bricks," she said, "will form the structure of great intersecting domes. Rap Rap has the plans. She will direct."

Rap Rap looked startled, but immediately began to draw up the plans. And in no time at all work on the structure was well under way.

*[You make it sound so easy. Drawing up the plans was hard work. Couldn't we have a chapter on architecture as well? Rap Rap]*

"Now," said Verity, beaming at everyone. "Any questions?"

It appeared that all the sisters were perfectly satisfied. No one had questions. Then Lady Shy raised her hand. She had been treated just like the others. She wished to make her mark. Now she said, "Please, what are we supposed to use for mortar?"

"The sands of the sea," Verity replied carelessly, "mixed with a little primeval mud."

"But the sands of the sea are white," cried Shy. "The sand won't hold. Crone Kronos requires blood!"

*[I have read your account. This is not what happened. The desert was not a desert. It had already been seeded. It was chock full of memes. And Babel wasn't built out of bricks alone. Babel had gardens, fields, meadows. Not sandstone alone, not machines and machinations. Where are the red and green colours of Babel? You have made me out to be a*

*bloodthirsty barbarian. But life and death, green
and red, actually matter. In the name of Crone
Kronos, in whom you appear to believe, where are
the Children of Babel? What did they eat? On what
did you feed them? Did they prosper? For the people,
Lady Shy]*

Cinders gasped. Lady Shy would have to be dealt
with. Verity was clearly out of her depth. She
adopted her most queenly manner. "What do you
mean?" she asked Shy.

But Shy was not going to be silenced so easily. "I
mean," she replied, and she raised her voice so that
the other sisters could hear what she was saying,
"Crone Kronos requires a blood sacrifice. Crone
Kronos requires that her doctrines be set forth
clearly. Crone Kronos requires a priestess."

"You seem to know a great deal about Crone
Kronos," Cinders murmured. She and Little Red
waited for The Black Piglet to say something. But
The Black Piglet merely looked quizzical.

Meanwhile, Lady Shy had begun haranguing the
multitude. "Which of you will sacrifice herself for
the greater good?" The sisters drew back, but they
couldn't help listening. "Shall a sister be sacrificed
that the sisters might live? How can that avail? No,
let a sister be found who is not a sister. That such a
one should die for the many is only fitting."

Upon hearing these words the sisters began peer-
ing into each other's faces. Which of the sisters was

not a sister? Then whispering began, "Men are not sisters. Let us sacrifice a man . . ." One sister, somewhat bolder than the others ventured to ask, "Are men sisters?"

Little Red and Cinders had to act. They could see where this was leading. "Of course men are sisters," Little Red told them in a strong voice. "All men are sisters, though there might be some men who haven't understood that that is what they clearly are." She winked and grinned and the sisters relaxed.

Lady Shy was reluctant to take on Cinders and Little Red if she didn't absolutely have to, so she joined in with them, "Of course, men are sisters." She paused. "But surely there is one among us who is not a sister by any definition." And with these words she looked straight at The Black Piglet.

The Piglet seemed unaware of her danger. She still looked bemused. But Solitude's voice rang with passion as she sprang to the defence of The Black Piglet. "The Black Piglet is my sister! And yours and yours, provided you are worthy and dare make that claim."

Again there was a murmuring among the sisters and a change of mood. They all wanted to be The Black Piglet's sister. Lady Shy had to regain the initiative. "And who are you?" she asked Solitude.

"I am Sister Solitude," Solitude replied with immense dignity, "and I know more about Crone

Kronos than you are able to imagine. Only The Black Piglet knows more. But you, Lady Shy, have committed heresy. Crone Kronos is angry with you. Any more nonsense and we will cast you out!"

Solitude was so furious she didn't really care what she said. It worked nevertheless. Lady Shy subsided, and appeared chastened, even penitent. She decided she had better placate Solitude. And anyway she had got what she wanted: it was clearly recognised that she was someone. She went down on her knees. "O High Priestess of Crone Kronos. I am an unworthy acolyte. I ask your pardon."

Solitude turned away. This was not the right time to say what she thought – Cinders and Little Red were persuading the sisters to get back to work – but she swore to herself that whatever she was or might become, High Priestess of Crone Kronos she would not be.

*[Shy: How very noble! A more partisan account would be hard to come by. You haven't the faintest idea what I'm really about, have you? The motives you've ascribed to me seem to be a composite of personal ambition and mindless blood lust. Babel wasn't built in a day, you know! Nor did it rise to your tuneful tale-tellings. It was built by the sweat and blood of women, a multitude of women! But these remain nameless. The mindless mob. They eat what they're given, swallow and digest – as best they can – whichever tall tale you choose to tell them. And by*

*whose authority? An invented Crone Kronos! Some
sort of monstrous, mysterious Matriarch! Where was
Crone Kronos when the sisterhood spoke of the need
for Babel? It is we who make Babel, and it is for us
that the sisterhood dreamed of Babel. But you! You
are murderous. You would wilfully obliterate us all!*

*Sol: But Crone Kronos exists! Time is a mystery,
present and palpable. It is there. And here. It is anywhere.*

*Shy: Time is the blood that runs through our
veins. While we laboured, you and The Piglet sat
back and talked! Don't pretend you don't understand. In the climate of Crone Kronos there are the
Eaters and the Eaten – your own words, Solitude. I
spoke for my sisters. I spoke for justice. I spoke for
the powerless against those in power!*

*Sol: You spoke for yourself!*

*Shy: And why not? Do you think that people like
me don't matter? When you cut us, we bleed. Don't
be so squeamish. Call me by name. I am not Shy. I am
Shylock! Lady Shylock! Part of the dust and destiny
of Babel!]*

That evening, after the toiling sisters had gone to
bed, or gone to a tavern, Cinders, Rap Rap and
Little Red held a meeting of the Inner Council,
though they'd have denied that such a body existed.
Solitude and The Black Piglet were both present.
Verity was very much present and Charity was

unobtrusively present. She didn't say anything, just busied herself with serving the others. The cat, having been cajoled back, was intermittently present.

Rap Rap chaired. Her approach was direct. "Look," she said to Solitude and The Piglet, "we didn't really want to contradict you in public, but what's all this about Crone Kronos? Why are we worshipping her? Is she a goddess? Is she a relative? Does she exist? Is she your aunt?"

The Black Piglet didn't quite know how to offer a precise answer. Solitude was frowning and shaking her head. But Verity flung herself into the breach again.

"Crone Kronos," she said, "is the mother of everything, and the grandmother and the great-grandmother. She is in everything and of everything, and without her there is nothing."

"Yes, I see." Little Red felt puzzled, but she said comfortably, "Should we build her an altar then? How is she worshipped? You see, it all has to be organised and set up."

"And what is orthodox?" Cinders enquired. "What exactly do we tell the sisters? We'll have to tell them something. For example, is there such a thing as a sin against Crone Kronos? What are her commandments? What does she want? And in return what does she give? Or doesn't she give anything? But then what are her principles? Her philosophy? Her theology? Who are her relatives? We can't just say 'Crone Kronos, Crone Kronos', over and over. What is the Catechism of Crone Kronos?"

The cat's ears had pricked up at the sound of her name, but The Black Piglet just looked blank. Cinders turned to Solitude. "Solitude, will you be the Priestess of Crone Kronos? At least tell us what she looks like. We need an image."

"Use your own image," Solitude replied. "She looks like you, like me, like anybody. That way she'll at least look like somebody. But as for being High Priestess, no, definitely not."

*[The post that you wanted, Lady Shy, I turned it down. The Piglet and I – unlike you – were not concerned with prestige and power. Sol.*

*In that you're to blame. You were irresponsible. Shy]*

"Why? Aren't you a servant of Crone Kronos?"

"Everyone is."

"But aren't you an especially good servant?"

Solitude shrugged, but Cinders persisted.

"Well, then who among us best serve Crone Kronos?"

"The artists and poets."

Cinders sighed. These were precisely the most unreliable of the sisters. "We can't have an hierophancy made up of them. They're the hardest to organise. They think about other things. They make up things. They wander off."

"Yes, they make up the tales of Crone Kronos."

"Why?"

"That's their function."

"And what's ours?"

Solitude turned to The Black Piglet, and The Piglet said, "To find a language that either captures or co-operates with Crone Kronos, that like Crone Kronos is metamorphic, and that does not lock us into individual cells."

The others didn't know what to make of this answer. They tried to ask more specific questions.

"Where does Crone Kronos live?" asked Little Red.

"In our BRA-INS!" replied Verity with evident relish. The Piglet and Solitude did not contradict her.

"Well, who is Crone Kronos? What is her identity?" Rap Rap ventured.

"That's just the problem," Solitude told her. "Identity is fluid."

Cinders and Little Red were ready to give up, and Rap Rap lost patience.

*[Solly, Shy can obviously be a pain at times, but she does have a point. If only you and The Piglet had been a little more practical! Little Red]*

"Look," she said, "just give us a few simple rules which the sisters can practise in order to express their devotion to Crone Kronos."

But this was greeted by silence, until at last the cat spoke: *"Rule One: Everyone at all times to do their utmost to speak in poetry."*

"Do you mean in verse?" murmured Cinders uncertainly.

"No, poetry." The cat was adamant. *"Rule Two. No noun ever to be used unless accompanied by a verb."*

"What do you mean?" asked Little Red

"I mean what I say," replied the cat. "You know – like the lounge bars of certain hotels – no men allowed without ladies."

"I see."

"Yes, that's a good sentence." The cat nodded grave approval. *"Rule Three: the verb 'to be' to be proscribed entirely."*

"You just used it!" Verity accused. The cat ignored her. "And the penalty for breaking any of these rules: the continuous use of the continuous present for a whole week."

"What?"

"Surely you are knowing what I am meaning," the cat explained (unhindered by any trace of liberal guilt), "– Indian English. He/she/it/they will always be doing; they will never do, whatever it was they might have been doing." The cat grinned. "Will that do to be going on with?"

Rap Rap, Cinders and Little Red sighed. What about a Catechism?" Verity demanded.

"This conversation. Use the minutes of this conversation," the cat replied.

*[Look if you must mix with metaphors, tell them clearly that the bricks and the machines did absolutely nothing all by themselves. It was the memes, the memes buried in the sands of Babel, that really mattered. Don't you see? Some memes lived, mutated and grew. And some died. Yours sincerely, The Cat]*

### 3

The next morning the sisters and the workers and the fellow bricklayers, led by Lady Shy, stormed the apartments reserved for The Black Piglet. They were wielding their pickaxes, their shovels and their trowels, and they demanded a hearing.

"The silence of The Black Piglet is insulting to us!"

"The tales of The Black Piglet are too hard for us."

"The Black Piglet is élitist!"

"The Black Piglet is a pig!"

"The Black Piglet is an alien aristocrat!"

"WE WANT—

A religion of the people!"

"WE WANT—

Cause and effect!"

"WE WANT—

The Black Piglet's Death!"

The conclusion of their logic surprised many of the women, but before they could pause and think, Lady Shy began a new chant.

"The Black Piglet has blasphemed against the goddess.

She has called Crone Kronos –

a bad-tempered mistress!

She has called Crone Kronos –

a capricious companion!

an unsisterly sister!

a – a – a – kind of caterpillar!"

Lady Shy filled her lungs again. "The Black Piglet has sinned against Charity! Against Faith, Hope and Identity!"

"She hasn't sinned against me," Charity piped up, but nobody heard her.

Shy continued. "Crone Kronos must be placated. Crone Kronos must be sated. Crone Kronos requires the death of The Black Piglet!"

*[Shy: Solitude! I see what you're doing. You're intent on proving that I was intent on killing The Black Piglet. But can't you see that you are responsible? You've written it into the logic of your language! You and The Black Piglet wanted to sit on a sand dune and dream of Crone Kronos. And not content with dreaming, you broadcast your tales. But ideas into action alter and falter. You can't build Babel and keep your hands clean. MYTHS MUTATE!!!*

*Sol: Shy, The Black Piglet was never interested in power or politics.*

*Shy: That is my point!!!*

*Sol: What is your point?*

*Shy: That a dreamer on a battlefield is bound to get slaughtered. You are trying to make a martyr of The Black Piglet!*

*Sol: But neither The Black Piglet nor I did anything.*

*Shy: That's my point!]*

Rap Rap, Cinders and Little Red were horrified. They looked at The Piglet, but she appeared to be making notes on cadence and the logic of language. Cinders stepped forward.

"You are being absurd—" she began.

"You are being absurd!" mocked Shy.

"You are being absurd!" echoed the crowd.

Rap Rap and Little Red were disposed of similarly: Little Red on the grounds that she was excessively rich, and Rap Rap on the grounds that she had had an education. Just as Solitude was about to tell the Lady Shy and her followers what she thought of them, Verity pushed forward, shoved aside Solitude, and throwing her arms wide, called to the crowd: "LONG LIVE CRONE KRONOS!"

The crowd experienced joy and solidarity, a sense of oneness, a dream of happiness. "LONG LIVE CRONE KRONOS! LONG LIVE CRONE KRONOS!" They shouted back. The Black Piglet was doing her best to suppress a smile and had stepped back a little behind Solitude's back, but Verity dragged her to the front again. "Smile!" she hissed at her. The Piglet blinked, then attempted a smile.

"Do you know who this is?" Verity asked the crowd.

"Yes!" the crowd shouted. "Yes. She is our sacrifice. She is The Black Piglet."

And "Yes," Verity shouted. "Yes! She is The Chosen One. She has been chosen by Crone Kronos. She is Crone Kronos' Best Beloved! She is the Daughter of the Great Goddess!"

*[Now you will say, Solly, that I was telling lies, but that is exactly what I was not doing. The Test of Reality is The Test of Truth. Truth must be verified. For a truth to be true, it has to be corroborated! Verity]*

The crowd of women stepped back a little and gaped at The Black Piglet. Was she really the Daughter of Crone Kronos? She looked very like anyone else. But yes, there was something strange about the darkness of her eyes, about the set of her ears. The crowd gaped. "Don't contradict me," Verity hissed at The Black Piglet. "Why should I," replied The Black Piglet. "We are all daughters . . ." Verity wasn't listening. She had turned to address the crowd again.

"Are you true followers of Crone Kronos?"

"Yes!" roared the women.

"Then listen to the word of the great goddess, for she has sent her own daughter to give you this message."

Lady Shy could bear it no longer. "She is a false prophet!" she raged at the crowd. "Don't listen to her! Don't listen to The Piglet. The Piglet, after all, is only a piglet!"

But the crowd wasn't having any of this. The Daughter of Crone Kronos was far more interesting than any ordinary piglet. "Blasphemy!" they roared.

"Now fall on your knees!" Verity commanded. To The Black Piglet she whispered, "For Kronos' sake and your own, say something wonderful!"

The Black Piglet spoke. "The Three Laws of the Mother of all Mothers, the Ruler of the Universe, the Embodiment of All Being, are OBEDIENCE, PRAISE and HAPPINESS. You will obey Crone Kronos. You will praise Crone Kronos. And you will be happy in the worship of Crone Kronos. That is all for today."

The women raised their faces to The Black Piglet. They sighed and swayed. They had been given a taste of ecstasy, they had been given something they thought they could understand. They were content. They allowed themselves to be led back to work. Lady Shy was locked up securely in a remote chamber. The crisis had been averted.

Rap Rap, Cinders and Little Red were relieved; Verity was pleased, especially with herself; and Solitude was discomfited. "How could you?" she said to The Black Piglet. "How could you?"

But The Piglet's reply was mild. "I gave them a judicious mixture of what they must do, should do, and want to do. It was only common sense."

"What do you mean?"

"Well. We must obey Crone Kronos. There is no choice. We should praise Crone Kronos and her works. That, after all, is the function of poets. And we want to be happy. I took that for granted."

Solitude frowned, and then when The Black Piglet winked – outrageously, she thought – she didn't quite know what to do with herself.

*[Shy to Solly: You see what I mean. If you'd had your way . . .*

*Sol in reply: You'd have won right away?]*

For the next little while, work on Babel proceeded quickly. Domes and minarets reared in the air, reprographic machines were linked and set up, *[You do this in half a sentence. Don't you understand? It took a long time and it was hard work. Shy]* and the triumvirate, Cinders, Rap Rap and Little Red, considered the distribution of important portfolios. *[And you do this in the other half; but running a society is no joke. This too was hard work. Rap Rap, Little Red and Cinders]* The Piglet and Solitude opted out. Cinders offered herself as Head of State. She said she had had practice in being ornamental. Little Red wanted the Home Ministry. She was particularly concerned with the Question of Men, and with Cinders' backing had drafted a bill that decreed they were definitely sisters in name. Rap Rap had wanted the Ministry of Culture, but when Verity asked that she be put in charge of Reality and Religion and the Problem of Porn, Rap Rap had stepped aside, and begun toying with Finance instead. *[Toying! Really, Solitude, you do us all less than justice. Rap Rap]*

The days passed pleasantly enough. *[For whom? I had been incarcerated without even a trial! Lady Shy]* Solitude was busy thinking. The Piglet was busy scribbling. They woke up one morning to learn that Lady Shy had effected an escape, and had set herself up as the leader of a reformation. *[This was not easy. And I did not set myself up, I was elected. Shy]* It was rumoured that extracts from the diary of The Black Piglet had been copied and distributed, and that they provided ample proof that The Black Piglet was a reprobate. It was further rumoured that Lady Shylock was demanding blood or a place in the cabinet or possibly both. Events made it clear that she was indeed demanding whatever she could get. *[I was demanding justice and something for the masses they could understand and digest. Shy]*

Rap Rap and the others were surrounded once again, and only released on the following conditions:

that Lady Shy be included in the leadership, preferably as High Priestess of Crone Kronos;

that Sister Solitude be prevented from ever opening her mouth;

that Sister Verity be sworn to allegiance;

and that The Black Piglet be charged with treason.

Cinders, Rap Rap, and Little Red felt bad about agreeing to any of this; but they told themselves they hadn't much choice and agreed in the end without too much fuss. *[Solly, you make those of us who were interested in politics, that is in making sure that we had a functioning society, seem very shoddy. But*

*Shy has a point. To have done nothing at all does not make you blameless. Rap Rap]*

On the day of the trial Lady Shy as High Priestess of Crone Kronos held up a six sided brick high in the air and read out what she said were the scribblings of The Black Piglet:

*March 21   Crone Kronos a vast stomach. The past her excreta.*

*June 23    Is Kronos an infection? A virus of some sort? But necessary. Without her we'd die. And with her? – We die more slowly.*

*June 24    But who is Crone Kronos: how shall we carve her image? Crone Kronos is a cat who eats her kittens.*

*July 1      Kronos doesn't just carve the rock face, she carves live flesh.*

*August 15  What do we burn on the altar of the Crone ? Cigarettes.*

*no date    Can't store money – not very well. It erodes or evaporates – devalues is the term. Have to store it like vegetables in a vegetable patch, i.e. some of it perishes and some of it grows. So then? Is Crone Kronos like money? Is Kronos money? Partially perishable, but much needed, much wanted, much vaunted. And entirely precious?*

Lady Shylock gazed at the assembled women. They did not appear sufficiently shocked. She would have to expound, to expatiate. "Consider what The Piglet has said about our goddess:

She has said that the works of Crone Kronos are so much shit.

She has said that Crone Kronos is a mortal disease.

She has said that the goddess devours her own.

She has said that the goddess is a type of pervert.

She has ridiculed the goddess and blown smoke in her face.

And sisters, she has said that the cult of the goddess is materialistic and that the goddess herself is composed of money.

Sisters! Shall such a creature live? Shall such a one be suffered to sin? No, such a one must be slaughtered that her blood might pay for the sins committed."

With that Lady Shy seized The Black Piglet and pressed a knife against her throat. The sisters watched in frozen horror and fascination. Nothing so exciting had happened for days. But Solitude, shook off her guards and rushed to the rescue of The Piglet.

"Stop!" she shouted, "Stop! The Black Piglet is not The Black Piglet. These drops of blood have metamorphosed her." She pointed to the knife Lady Shy was holding, it was beaded with blood. Solitude then supported The Black Piglet and ripped off the mask that hid her face. "Behold!" she cried, "The Black Piglet is in reality The Black Prince!" "Behold!" she went on, "the great goddess has sent her son to be your bridegroom!"

*[Now who's lying? While I was only getting at the truth in an orderly fashion and trying to offer some sort of trial. The way you've written it, I am a murderess and you are the saviour of The Black Piglet. But in the interests of Babel, Solly, ought The Black Piglet to have been saved? You would say she was too good for Babel. Precisely. The fault is excess, an excess of superfine dreams and discoveries, and a lack of the simple, the true, the definite. I am vindicated. Shy]*

The assembly of women examined The Pig-Prince. He made a fine figure; and though his visor was down he seemed comely. And when he spoke his voice was gentle, soft and low, an excellent thing in a man. "Which of you?" he said to them and he smiled on them equally, "which of you will be my bride?"

Lady Shy flung herself at the Prince's feet. "I, my lord. I as your High Priestess, I am the chosen one. I shall be your bride." *[And why not? It seemed to me then that The Piglet was not quite so unworldly as you, Solly. I seized my chance. I made an attempt. Shy]*

Cinders, Little Red and Rap Rap looked on in disdain. Were there no limits to Lady Shy's ambitions. *[And you say that Cinders et al are not élitist? All they tried to do was maintain the order in which they had a vested interest. Shy]* What would the Pig-Prince do now?

"No," he said softly. "No," he said sadly to the prostrate Shy. "You are not worthy. You have sinned against my mother. You have sinned against me. You concocted these heresies yourself, Little Shy. Your mind is a cesspool. You must pay."

"Take her away," he commanded abruptly. Three or four women scurried to obey him. Then he turned to the others, "And you my sisters, think upon the mystery you have witnessed today, and pray that you too might live in the mercy of my mother from moment to moment." *[And The Black Piglet now, sending me away – more in sorrow than in anger, it is true – is she as blameless as you make her out to be? A villain by any other name is still a villain. My only crime was that I was relatively powerless. Unabashed and ill-understood, Shy the Blameless]*

The counter-coup was over. Rap Rap and the others assumed once again the reins of power, albeit a little shakily. *[Were your hands steady, Sol? Pace, Cinders]* The question was what was to be done with The Black Prince? Was he still The Black Piglet? And if not, why not? And who had been lying to whom? And when? They retired to the Inner Chamber to sort out once again, who was who and which particular chairs belonged to them.

The Black P seemed to be behaving as usual, but Rap Rap was having trouble with her eyes. The

Black P's image kept flickering, from piglet to prince and sometimes to princess. The really serious difficulty was that the pattern and meaning that everything else made flickered as well and re-arranged itself.

"Look, could you be consistent?" Rap Rap said crossly.

"Consistent with what?"

"With yourself."

"I probably am." The princely piglet smiled. "Do you know, I was thinking, since the existence of the present is contingent on the extinction of the past, it follows that from moment to moment Crone Kronos dies simply in order to create herself."

"Don't talk nonsense! We have a crisis on our hands."

"What crisis?"

"The fact of your flickering!"

"But that was just what I was explaining. If I had been startled into seeing you as you see me, you'd be flickering too. Everything is flickering. It's just that it's not comfortable to see it like that."

"It certainly isn't!"

Little Red intervened. "Tell me," she said. "Why 'prince' why not 'The Black Princess'?"

"Do you know, that was just what I was wondering. Why not, for example, The Black Parakeet or The Black Peony? You see, it seems unreasonable to me that there should be only three distinctions of gender and two of number (usually). Why not distinguish according to rank or species or colour? For example when I speak as The Black Piglet, I could say 'i' rather than 'I'. And when Solitude and I speak for ourselves and are in agreement we could say 'We' instead of 'we'. But when we speak for the whole of Babel perhaps we should say 'One' and speak of ourselves in the third person as a means of arriving at a superlative plural?"

"Please!" Cinders said softly. It was obvious to her that in her present mood The Black Pig-Prince was not going to be helpful. She turned to Solitude. "Do you understand what is happening? How do we explain this most recent marvel? And what will it mean for the Building of Babel?"

Solitude paused, then she said, "The Black Princess' logic is completely in keeping with what happened to Babel: first consensus – about building Babel; and then, the inability to understand any individual language."

Little Red ignored the latter half of Solitude's speech. "How do you know that The Piglet is now The Black Princess?"

"I know what I choose to know," Solitude replied. "Look, we can all insist that everyone and everything essentially is what we agree it is, or we can learn to live with the metamorphosis of The Black Piglet. Don't you understand? That is the challenge. That is the test."

"The test of what?" asked Little Red.

"Of our intelligence," said Rap Rap.

"Of our goodwill," said Charity.

"Of our courage," said Cinders.

"Of our integrity," said Verity.

"No, no, no," Solitude was almost shouting in exasperation. "It is the Test of Crone Kronos, of the goddess who dies and invents herself. The Black Piglet both is and is not. She, like everyone else, is suffering metamorphosis. You must find a language to describe that process, or else–"

"Or else what?"

"Or else empires will rise and decay and die, and rise again, and you will play at musical chairs, and Crone Kronos will have her way."

*[A dire prophecy. And all very beautiful and metaphorical. But the fact is, Solly, The Black Piglet was dying. If The Piglet was indeed in a special relation to Crone Kronos, then how explain this to the sisters of Babel? Perhaps you don't care, but there are those of us here who would like to be constructive. Would it be useful to say that The Black Piglet was an incarnation of Crone Kronos? Rap Rap]*

"And what's that supposed to mean?" asked Little Red crossly.

"It means," replied The Pig-Princess, "that unless you can fashion a tool that will trap Crone Kronos, or at least keep up with her, Babel will fall as it fell before and all your work will have been in vain."

Little Red was puzzled. "But are we supposed to trap Crone Kronos or worship her?" She looked at the Pig-Prince suspiciously. "Did you really write all that stuff that Shy read out? Are you an enemy of Crone Kronos?"

The Black Pig-Princeling looked embarrassed. "Yes, I did write that, but that doesn't mean I'm an enemy of Crone Kronos. Look, let me describe the problem in a different way. All quests are a quest for Crone Kronos. Think of her as an antelope, a beautiful antelope, whom we, her devotees, pursue."

"Is she fleet, this antelope" asked Cinders, doing her best to understand.

Pigling looked at Solitude in despair, and Solitude intervened. "Well, the thing is she re-creates herself from moment to moment. It's like trying to catch an antelope in a Walt Disney frame. Each time you try, you're in a new frame."

"With a new antelope?" Rap Rap inquired.

"Yes. And in the new frame a new questor as well."

*[I'm getting sick of this! Crone Kronos = antelope = caterpillar = Piglet's Mum. And Piglet = prince and/or princess = The Black Peony = The Black Parakeet. And at the same time Capital C Cat proscribes the verb "to be". It's metaphor gone mad! Verity*

*I think you've hit on something. Metaphor in motion is metamorphosis. Crone Kronos' doing. Solly.*

83

*And for Crone Kronos' sake let's call The Piglet The Piglet. We're used to it. And at least we'll know exactly what we mean. Verity*

*Will we? Solly]*

Rap Rap lost patience. "Look, we have a temple to build, a society to govern and a civilisation to construct. We can't tell the sisters all this rubbish about an intermittent goddess, and then tell them that they themselves are intermittent as well!"

"That's right," put in Verity. "The sisters need to be able to distinguish theology from mythology, tall tales from Truth, the canon from the apocrypha. Now, in all the tales that have been proffered lately, wherein lies the grain of truth?"

"In the heretical ones," murmured Piglet and Solitude miserably. "The tales chip away to reveal the process."

Rap Rap had had enough. She glared at everyone and adjourned the meeting. Just then, the cat who had put in an appearance, winked at the women, thus irritating Rap Rap further. By now Rap Rap had a severe headache which she blamed on The Piglet and Solitude equally. "At least when you reappear, you have the grace to be the same cat," she flung at the cat.

"How do you know?" retorted the cat. "Oh! Look out!"

They looked. Whether as a piglet or as a princeling, The Black Piglet had bled profusely. Her head fell forward, no more flickering: The Piglet was dead.

*[Shy: This is grossly unfair! You're suggesting that The Piglet died from that little nick on her throat when I was holding her. There would have been rivers of blood – all over the table.*

*Rap Rap: We would have noticed, Solly, whatever you think.*

*Shy: Set the record straight. I am not culpable.*

*Verity: Well, but who is then? Why did The Piglet die? Who killed her?*

*Shy: I told you, Death. Death killed her. It's inherent in all the stories they've told. Their own death, and the ultimate death of Babel.*

*Verity: But what does it mean?*

*Cinders: It means that we'll have to modify The Tales of The Black Piglet, and find someone as skilled as The Piglet in dealing with Crone Kronos, but it will have to be someone who's a little more practical.*

*A little more powerful.*

*A little more political.*

*Popular, plausible, presentable . . . Stop it. Please stop. Sol]*


How to explain the death of The Piglet?

The following day Verity was ready with an explanation. She sounded convincing, forthright and definite. "She is gone," she told the sisters. "Her blood has been shed. She is gone to her mother, for we are unworthy. But when we are worthy, she has promised to return, and in her retinue she will have a thousand bridegrooms, each one comely and bejewelled like the firmament." The sisters sighed, then applauded.

## The End

*[As required by Rap Rap I have printed out my notes on **The Life and Death of The Black Piglet**. I wonder what Rap Rap will make of them? What Verity and Shy will do with them? I should finish the account, and write myself out of Babel. Solitude]*

*[Really Solitude, what did you expect us to make of your notes? Rap Rap]*

Minarets and domes spiralled upwards. Babel rose, and eventually fell. The Pig, The Prince and The Princess were powerful symbols. In varying modes and manifestations each, at different times, governed Babel. Solitude was not mentioned either then or later, and does not figure in the building of Babel.

*[No, Solly. This won't do. We must prepare ourselves for growth and success, not resign ourselves to an inevitable end. Rap Rap]*

In a shorter time span the founders of Babel lived out their lives and entered history. Rap Rap grew increasingly irritable. *[I am known for the evenness of my temper! Rap Rap]* She is known as the author of a concise chronicle called "A Brief Biog. of The Black Piglet". This became a classic. For several hundred years the memory of Rap Rap was venerated. Then it was not. Such things happen in the course of Babel. Lady Shy achieved fame, particularly after her assassination. *[Why wish this on me? Shy]* This was attributed to a group of zealots who felt that The Teachings of The Piglet had been sadly distorted. Soon afterwards they were killed themselves. And all this was done in the name of Crone Kronos, a fact, of which Little Red who happened to be Home Minister, was not unaware, but which information she kept to herself. *[Oh for Crone's sake, Solly. Don't be so smug! Red]* Such things happen in the course of Babel. Cinders achieved a modest success, Verity was given a golden handshake, and Charity died of a wasting disease. There were moments in their lives when each was happy; there were moments also when they experienced regret.

As for Solitude, though not incapable of a historical perspective, when The Piglet died, late that night she threaded her way out of Babel. As she walked, her friend was a shadow walking beside her. Later, there was no shadow, only a trace of a trace, a fading presence.

The Very End

*[Solly, this is so self indulgent, not to say irrespon-
sible. You've left us behind to pick up the pieces with
at least half a dozen questions still unanswered. Rap
Rap, Cinders and Little Red*

*Look, you can do what you like – or what you
must. I have said what I have seen – well, most of it,
some of it. Any interpolations, any extrapolations are
not my business. My duty is clear: I must mourn my
friend. However, being conscientious, I am now
offering some final comments which, if you choose,
you may append. Solitude]*

## Appendix A
## Solly's Explanation

I, Solitude, witnessed the death of The Black Piglet.
She died because of Babel. No, not because she was
a pig. It was the sharing and the roaring. She was
neither a pig nor a princess. She was an ordinary
woman of unusual intelligence: brown skin, brown
eyes, dark hair. *[Both "unusual" and "ordinary"?
Perhaps it's you who undervalue the worth of your
friend?]*

Was The Black Piglet in love with Crone
Kronos? Yes, of course, she was. She was deeply in
love with all three of them. And which of the sisters
fascinated her most? That was the question I had
hoped you would ask. She was an avowed admirer
of Crone Kronos of course. She professed, and

indeed it was genuine, or at least, she wished it to be, she professed respect for Madame Mem, though in her heart it is possible that Mme Mem's frailties and Mme Mem's follies detracted somewhat from whatever it was Piglet felt. The sister – and this is the admission I did not wish to make – that the Piglet knew best – was Sister Death. *[Solly! Don't you understand? This is your interpretation, your "favourite" sister, your understanding of what happened in the end.]*

Sometimes, in the desert, when we were working together, Piglet would look over her shoulder. What do you see? I asked her once. "A shadow of a shadow. A figment." An enigmatic answer. I thought nothing of it. But sometimes in the desert her nostrils would twitch. What do you smell? "I smell burnt flesh." Roast meat, I thought. The Piglet is hungry. I thought nothing else. But later when Shy was thirsting for blood, I heard Piglet sigh, "She needs me, you know. She needs my flesh." *[You have invented the whole thing! A false fabrication! An unsubstantiated lie!]*

So then, was The Piglet for whom I've claimed heroic stature, less than heroic? She was afraid of Death. But that's common sense. Did you expect The Piglet to behave like a goddess? By definition mortals and immortals belong to different species. Why would a goddess want to bother with Babel? *[But perhaps she was a goddess, who did, indeed, wish to bother with Babel?]* The Piglet could hear Death's footsteps. And so? Was the quest for Crone Kronos merely an escape? Were the tales of Crone Kronos merely a noise, something to cover up the snufflings of Death? No, that is less than a fair judgement. She met her Death.

*Solly! Each of us can achieve that. It does not require a woman of unusual and acute intelligence. Surely that was not the end of her quest? O Sol, Solly, O Solipsistic One, whatever you've seen or think you've seen is subject to change – even your text!*

Read the text, you bloody fools! And that is my last word.

Solitude in Retirement.

# III

## *INTERLUDE*
## *WITH SOLITUDE FOR COMPANY*

*Does she scent death?*
*Does she scent danger?*
*Has she experienced her own anger?*

In her shack in the desert Solitude told herself she
was used to it, she wasn't lonely, but, as was not
unnatural, she missed her friend. She told herself
that if therefore she could resurrect The Piglet, she
would be doing her friend an enormous favour. The
Piglet would be pleased. And the quest they had
been engaged in would come to an end. With delib-
eration she began to construct a credible past in
order to deal with an insistent present.

Suppose I tell stories about you, present you, so
to speak, in your rightful image?
When Piglet was a baby she strangled two snakes.
These snakes were emissaries of Crone Kronos. She
knew that some day Piglet would be a hero, with
whom she, Crone Kronos, would have to contend.

Am I doing what the sisters did? Inventing The Piglet to solace myself? I would rather we were both alive and inventing each other, or inventing the scenery, or playing our old game and inventing Crone Kronos. But I don't think we have invented her.

The sisters could say: "The Piglet is dead." I could say: "Long live the Piglet!" Am I just as stupid as any of them?

I will tell you another story:

Once upon a time there was Prince Piglet. And then there wasn't.

Once upon a time there was Prince Piglet, and Piglet had a destiny. But doesn't everyone? Doesn't anyone?

Once upon a time there was a princess called Piglet, and I singled her out, she was my friend. That is the point.

There's metamorphosis: I followed Piglet to a greener land, a different country. There was water and there were trees. Piglet had turned into a green parakeet. She fed from my hand. But she wasn't Piglet.

Here's a story you might really like. One day Crone Kronos appeared to me in a long robe and a magician's top hat. "Guess what the future holds, my dear! Guess what I have in store for you!" She chortled merrily. Then in rapid succession she pulled out a white rabbit, a green parakeet and a black piglet, and all so delicate, so beautifully made. And then – who knows why – in rapid succession she smashed each of them.

Or there's the story of the miserable poet who went crawling up to Crone Kronos, saying, "CK, Hey CK, I'm half in love with easeful death," and of how Crone Kronos paid no attention at all to this broken worm – neither alive, nor quite dead.

Did we invent Crone Kronos? Verity believed that you had invented her. And I think that Rap Rap, in her heart of hearts, saw Crone Kronos as a kind of seamstress. Her function? To measure the distance between two events. She thought that the events were there anyway, that time was just the backdrop on which we happened to perceive them. But that would be like the embroidery saying to the tablecloth, "I made you!"

I have a question for you, Piglet. (Do you remember Verity's penchant for intelligent questions?) Does Crone Kronos mourn? Have you ever seen her cry? Perhaps you would say, "Yes. She cries sometimes." I might ask, "What does she cry for?" You would reply, "She cries for herself."

I will write elegies.
    I will select certain flowers,
precious moments culled from Crone Kronos
– beheaded flowers, but never mind that.
I will stick them in the sand.
"Look! Look!" I will say
                        to any passerby
"I have made a garden
            in the teeth of the sea,
                        in the arid sand."
And I will forget
            I had meant to mourn.


To whom will I say this?


Let's cut the cackle. You are dead and I am
dying, and in your death I have a vested interest.


Suppose Crone Kronos appeared to me? Suppose
she manifested herself in the guise of an old washerwoman with a laundry basket. (So as not to frighten
me too much. And presumably I could find sym-
bolism there.) Suppose she said, "All right. I'm here
now. What do you want?"

I could say, "I want The Piglet." But the whole
world would tumble and crack. Huge lumps of plaster would fall down, and behind the plaster there
would be black holes. And the plaster would look
like a crazy jigsaw.

I could say, "I want power over you."

"Sleep," she would say. "Sleep and dream."

"But I can't control my dreams!"

And perhaps she would smile, perhaps it would be a toothless smile, "Ah well."

I could say. "I want to BE you."

"Sure," she would say, "sure, sweetie. Come a little closer."

Even suppose we could catch the antelope, or the caterpillar or the washerwoman, what could we possibly do with them? What could we say?

Suppose I said (to the crazy washerwoman), "I want to be your partner."

What would she do? Dump time's laundry, all the soiled laundry that there ever was, in my unwilling lap. The smell! The smell of the diapers, the kitchen waste, of the world gone to rot, of the fruitful and fetid, of fungus! The smell of process! The smell of CK's body, of Mem's step-sister, triggering memories of time itself. Proust would have a ball. But me? You? I don't think I'm up to that.

Perhaps I could say to Crone Kronos, "I surrender. I am your devotee, your disciple."

And Crone Kronos? She'd just laugh.

The truth is, Piglet, at one point I went down on my knees to the old washerwoman. I told her she was a goddess and I prayed to her. "Bring me back all the green world when Piglet and I walked together in the gardens of Babel.

Bring me back history. Bring me back buildings, bring back the people, the sullen and stupid, the sweet and the sour, bring back everything, so that I might see Piglet again."

But Crone Kronos said it was the wrong prayer.

I could relapse into silence. Would that not make a pretty picture? Solitude Sinking Into Silence? How would one paint it? Blackness upon blackness?

And you? Piglet taking the form and colour of air. Fading into light. We are both transformed. But who are we then?

If we stop talking, who are we Piglet?

Shall we allow ourselves to suffer a sea-change to the point where we do not recognise ourselves?

Piglet, I've just realised. This is the proper version of the story you told:

One day when Crone Kronos and Madame Mem were going for a walk they persuaded their youngest sister to come along with them. She was inclined to be sulky and querulous, to skulk in the dark, to be flighty and foolish, to change her mind at the drop of a hat or a leaf or a feather, to be extremely short-tempered, and what is more, to be malicious and envious. She envied Crone Kronos and Madame Mem. They did what they liked. They ranged abroad. They pointed out that she too did exactly as she pleased, that she ranged abroad. But she said that it wasn't the same thing, that they appeared to enjoy themselves. "I've seen you licking your chops, little sister. Don't tell me you're not capable of enjoying yourself." Crone Kronos was inclined to be sharp with her. Mem, on the other hand, was generally patient, in part because she felt sorry for her, and in part because she was afraid of her. The truth was that their youngest sister was unusually stupid. There was only one thing she could do, but that she did supremely well.

Solitude frowned, and began again.

No, wrong story, Piglet. How it really went. Once upon a time there were two sisters, Crone Kronos and Death; and one had everything and the other had nothing. The one who had nothing envied the one who had everything, and she did her best to destroy what she could of the other's wealth. They also had a third sister, Madame Mem, who was not envious because she thought she had a little of everything and a little of nothing; but Death despised her since what she had or didn't have was all in her head. And so Crone Kronos walked and Death followed after and that's how it went.

Solitude considered and reconsidered this version. By this time she was no longer sure whether to attribute it to The Piglet or to herself. She decided to disown it.

In this story, Piglet, you make Crone Kronos heroic. You say she is Life and her opposite is Death. But is that true?

No, the real story is different. Mem has the brains. Yes, it's true, Death is stupid. But as for Crone Kronos she is BRAINLESS!!!

Solitude waited. Surely she had blasphemed loudly and sufficiently? She tried again.

No, I've thought about it, Piglet. Crone Kronos, Mem and Death have nothing at all to do with Reality. Reality is an altogether different kettle of fish. And Crone Kronos, Mem and Death aren't people. They're just screens or openings through which Reality flows. Crone Kronos' screen is completely transparent: Reality flows and nothing is lost and nothing retained. Death's screen is an ordinary wall. Everything's smashed, nothing is retained and nothing gets through. And Madame Mem's is a faulty filter. That's how it is. The three sisters are just screens.

And where are they located? In our brains. We eat time. And when we lose our capacity for eating, we die. Does it follow then that the more brainless we are, the more we survive? Because we eat less time? Is the death of a highly complex being much more serious than that of a stupid one? It is more noticeable. I, Solitude, am highly noticeable in an unpeopled landscape. But there's no one to see me.

Solitude waited. Still no answer. In desperation she pressed hard on her computer keys and asked for help:

WHERE ARE YOU PIGLET?

In no time at all a reply appeared:

*To Solitude*

*solitude@bp.des.bl*

*Don't be silly, Solly. All forgiven. Please return. the Black Piglet lives in the bones of Babel. You won't find her in the desert all by yourself. Queen Alice's coronation in three day's time. Need your help. For old time's sake, Rap Rap*

Solitude read and re-read this message. Then she typed back.

Maybe you're right. Maybe it's genes, it's DNA, it's life, it's energy, it's matter, it's matter that's important?

*No, not genes, memes. You are a highly organised individual chock full of memes. And so was The Piglet. Her memes are reproducing in the bricks of Babel, in the brains of Babel, in all the mechanisms and machinery of Babel.*

Does repetition matter?

*That's how the memes know they are sisters.*

But who are the memes?

*Memes daughters of Memory. That's why that chap named them so [Richard Dawkins].*

Hundreds and thousands and billions of memes,

flitting about like speedy electrons, each capable of reproducing and linking with her friends?

*Provided there's a host, a host of hosts.* I had better check out what's been happening in Babel.

*Memes mutate.*

But shouldn't The Black Piglet's memes be kept in their purity?

*Can't be done. The memes are in motion.*

And a meme without a host is of no consequence. But as soon as there's a host?

*The meme just whizzes into a different pattern.*

Who'd have thought that Ma'am Mem had so much energy in her? Or that she was so powerful?

*Or such a profligate.*

Listen, are we a conglomerate of linked up memes inhabiting the body of Crone Kronos? And is Crone Kronos a three-headed goddess?

*I don't know. When are you coming back?*

Yes, okay, I'll return shortly, but I have to write my elegy first. May our memes match! In sisterhood, Solitude

PS. Who the hell is Alice? And why are you crowning her?

*Don't you remember Alice? She has agreed to come and govern Babel. It was either her or Lady Shy. Alice is claiming that she is the Rep. of Crone Kronos on earth. Babel is changing too fast for me. Do come and help. Love, Rap Rap*

*PS. Why are you writing an elegy?*

Why shouldn't I? It will all help to build Babel –
that is, of course, when I return.

The trees were like plumes in Babel.
The river fed everyone and everything.
In Babel friendship was easy.
     A city rose out of nothing.

She strode through the fields of Babel.
Her breath made words that were pleasing.
In Babel the birds of prey lived off the air.
     The women were laughing and loving.

She smiled on the women of Babel.
They built her a temple to time–

This is all I've done so far. What do you think?
                                        Love, Solly
*But Solly it doesn't scan! Anapaests don't suit The
Piglet. Besides, Solly, it wasn't like that. Do come
back. We need to talk.*

                                     *Love, Rap Rap*
Well, then write your own elegies, do! Some sort
of realistic rubbish in Lady Shy's vein – is that what
you want?

Piglet is made of blood and bones.
The butcher's knife has hacked her up.
How shall these bones prevail?
Oh come and dine, eat her up.
I won't even put my name to this.

*Don't be cross, Solly. It's not Lady Shy we have to worry about. It's Queen Alice. She wants to let men into Babel. She says men, women, dogs, cats, piglets – it's all a matter of rank. She wants to rationalise the ranking system. I'm seriously worried.*

*And I do see the point of your elegy. Things really were much better then.*

*With love, Rap Rap*

I don't think I much like the sound of Babel. Why don't you come here? Love, Solly

*You know why. It's lonely there. Love, Rap Rap*

Solitude wanted to cry and sing all by herself, and try to make her crying musical. But neither she nor The Piglet could thrive in the desert.

Sol, Solly, The Solipsistic One
Couldn't build Babel on her own.
Brain power's required,
hooked up and wired,
to change and alter what is known.

After a few more days of sighing and crying, she allowed Rap Rap to coax her back.

But before she left Solly told Crone Kronos exactly what she thought of her.

"I think you should be made to sit down and mourn for all your children."

Crone Kronos replied, *"Such a grief would*

*drown the universe."*

This baffled Solly, but she answered back, "Perhaps that's why you don't do it. Perhaps it's as well that you're so insouciant."

And to The Piglet Solly offered three speculations:

"Crone Kronos is what the brain produces when it imitates time."

*"And Mem?"*

"Mem is what the brain produces when it imitates its own process."

*"And Death?"*

"Death is what the brain produces when it is unable to produce anything else."

To which The Piglet replied: *"Nevertheless, I did not invent them."*

# IV

## *QUEEN ALICE*

### 1

For Queen Alice's reign I am the obvious, omnivorous narrator. Alice and I go back a long way. Am I her cat? No. Is she my human? Don't be sentimental. Alice and I know one another, but in the past our encounters were confined to chance. That is what makes me objective. Girls and gargoyles, gryphons and gorgons are all one to me. I've kept my distance and sharpened my claws.

Verity objected that as an observer I might be selective, if only because I appear at random. She assumes that if she cannot see, she can't be seen! That's truly subjective! But Verity no longer matters much. Rose Green and Snow White – Alice's stooges – have taken over. Rose Green's job: to decide which gene or meme shall live. Snow White's job: to snow over.

Queen Alice herself displayed annoyance when she discovered I was engaged in immortalising her. She said I wasn't one of her subjects, declared I hadn't a sense of humour, and pointed out I had little practice in dealing skillfully with royal phenomena. True enough. I've sworn no loyalty to Alice or to Babel. She isn't paying me. Not that she'd have considered it – paying me, I mean. Alice's idea is that other people ought to pay her for permitting them to live. A thug? No. Just Queen and Protectress. Anyhow, even without patronage, I've managed very well. My contract: only with you, the Reader, for whose sake I offer explanations and efface myself.

In order to understand Queen Alice's actions, her major motivations, her successes and dissatisfactions, it is necessary to distinguish between her and The Piglet, even though she frequently claimed that she was only carrying out The Piglet's plan. The Piglet was a dreamer, Alice an actor. She wanted to be something, to do something. She was an achiever. Solitude, whom Rap Rap had brought back, disliked her at once. Lady Shy was more resilient. And Little Red and Cinders, who were semi-retired, watched askance. Very shortly after Alice had taken over, Rap Rap was placed under house arrest. Alice herself was pragmatic as hell. She used whom she could, chose a direction, and went for it. She wished to make a notch in history, to be noted and notable, a flaming success. Alice was mad on immortality, which is why she could never quite bring herself to excise this document. One of her first acts was to establish a power base. This she did by simple decree: "All genes and memes belong to the Queen." It felt like a truism, and it followed logically that all her subjects ought to pay a tax. Alice's motto was easy to remember and soon appeared on all her banknotes: "Render unto Alice what is Alice's." Cinders *et al* soon regretted Alice's advent.

An invitation had been sent in the name of Rap Rap: would Alice care to succeed The Black Piglet? Possibly in the role of The Black Princess? Alice had replied that she wasn't interested in being a princess, it was Queen Alice or nothing. And in their desperation – after all, here was someone who seemed genuinely to understand power politics – they agreed to her terms. There would have to be a coronation. She would be the worthy successor of The Black Prince. Prince? Yes, Prince. Alice considered Piglet absurd. Why not Princess? Because, Alice faxed back patiently, though The Piglet had had the virtues of a woman – patience, kindness and lack of bossiness – to link these virtues with the power and prestige of a male persona would illuminate them. Unfortunately, in Babel, The Black Piglet was commonly known as The Black Piglet, and Alice, being practical, let it go at that. But there would have to be a coronation, a masque and some pageantry. She, Alice, would send her assistants, Snow White and Rose Green to help and direct. Had she made herself clear? Yes, she had.

Queen Alice was duly crowned. There was some pageantry. Everyone was dressed in green and red, which Alice had declared were the colours of Babel.

And there was a masque. It was called The Masque of Mem and her Sisters. It was widely rumoured that Queen Alice had condescended to write it herself. Lady Shy was Madame Mem, and Snow White and Rose Green were Crone Kronos and Death. All I recall is that Mem slew Death (Rosie clad in Sherwood green, slashed with scarlet), while Crone Kronos, that is Snow White, clad in something diaphanous, just stood about and looked helpless. There was applause. Queen Alice graciously accepted Mem's homage.

Then the Queen addressed her subjects: "O women of Babel you have sanctified me. I, successor and descendant of The Black Piglet, am now Crone Kronos' representative in Babel."

The women gaped. This was news to them, but it seemed like good stuff. There was more. "O women of Babel, we have built worthily, we have built well, the genes and memes of Babel proliferate. I am their guardian. Let those who would serve me swear their fealty and their faith." Well, even though Lady Shy was getting on a bit, she was still spry, and before the others had finished blinking, she was right there at Alice's feet. Alice then said, "Henceforth you will be known as Lord Shy, Keeper of the Memes." Shy was startled. What was wrong with being legitimised as Lady Shy? But she didn't demur; it was not her place. Alice then handed out a few more titles: Verity's daughter became Lady Verisimilitude,

Rose Green and Snow White were each given duchies i.e. a share of the revenue from memes and genes, and Rap Rap and the others were duly commended – for past services. Queen Alice rose, and before exiting, she smiled on her subjects, declared the day a holiday and told them all to enjoy themselves.

Rap Rap, Cinders, Verity and Red didn't know whether to be horrified or mystified. Solitude had her doubts, but went along when the others asked to see the Queen herself. Charity stayed at home; it was clear to her she had no part in the new government. They were given an appointment, and when at last the day came, five of them trooped to the palace. At the door to the Council Chamber Lady Verisimilitude, Verity's daughter. cautioned them. "Now, don't forget. Only sit when you're told to sit, and leave backwards." When it looked as though Verity was about to protest, she added hastily, "Now, mother, it's nothing to worry about. Just a matter of appearances."

They went in. Queen Alice was charming to them; she asked them to sit down right away. Lord Shy was seated on her right hand, and the two duchesses on her left. Before Rap Rap could say a word, Queen Alice turned to her. "We have just settled that the tax on the use of memes and genes only applies when memes and/or genes are used in public. Their use in private – until further notice – is unrestricted."

"I don't quite understand," Rap Rap murmured.

"It means," Lord Shy replied grandly, "that when memes are used in public, words for example, a tax must be paid. And similarly when children are created, who will eventually claim citizenship of Babel, a licence must be applied for and their dues deposited."

Solitude had resolved to show her disapproval by maintaining silence, but before she knew it she found herself saying, "And the Queen's speeches, are they taxed as well?"

Rose Green interposed quickly, "It would be meaningless for the Queen to tax herself. The Queen's words are tax free and are to be seen as a gift to Babel."

Alice frowned. "Come, let's get to the point gentlemen. I have asked you here–"

Rap Rap interrupted, "We are not gentlemen!"

Snow White rose. "The Queen must never, you understand, NEVER, be interrupted."

And Rose Green added, "OR CONTRADICTED" Alice waved them down. "It's all right. Rap Rap has the privilege of age. But that is precisely my point. I have asked you here in order to discuss The Question of Men, even though, just at present, there is only one man present."

They all looked at Shy, who smirked self consciously. Had Alice gone mad? Verity opened her mouth to speak, but Rose Green got in first. She said smoothly, "Consider the words of The Piglet. Why one gender or two or three? Why not hundreds? It is all a matter of rank after all. Gender is not immutable. And progress through the ranks is always possible."

"Which is higher? Men or women?" Little Red blurted out.

"I am not much interested in men or in women," Alice replied grandly. She decided to use the plural. "We are interested in order. In general, the noble outrank the ignoble."

"Like a game of cards!" Solly thought, but she kept this to herself.

"I am of the opinion – again in general – that men outrank women," Queen Alice continued.

"But not in Babel!" Cinders, who really should have known better, had interrupted.

"NO INTERRUPTIONS!" Rose Green roared.

"NO CONTRADICTIONS!" Snow White chimed.

Alice waved them down. "Let it go this time. Very well. In Babel women outrank men. Is that the consensus?" She turned to Rose Green. "Will the memes hold?"

Shy was about to make an objection, but Alice overrode her. "It's all right Lady Shy. You are who you were. Nothing is lost though your title has changed." Shy subsided, and Alice continued, "The highest rank of all, of course, is that of 'Piglet'. That is held by the Queen herself."

"Forgive me, Your Majesty," Cinders said quietly, remembering her manners, "but I haven't understood the point of all this."

Snow White rose to her feet slowly. "Power," she informed them, "the proper distribution of power, that is what makes for a stable realm."

"But didn't The Black Piglet say that everyone is a daughter of Crone Kronos?" Red asked doubtfully.

Alice nodded to Rose Green, and Rose Green explained. "I have scanned and sorted the words of The Piglet. And I can assure you beyond any shadow of a doubt that the word 'power' was never once used by The Black Piglet. In this matter as in all other matters we must now do what Queen Alice says."

"Why?" Solitude couldn't help herself.

Alice didn't explode. "Sister Solitude, to you much is forgiven, because it is rumoured you once associated with The Piglet herself. Times have changed. I make the rules. You obey."

"Why?" Solitude was pushing it, but Queen Alice continued patient.

"So that then we can have order, justice and fair play!"

Snow White and Shy applauded loudly, and Rose Green made a point of noting down what the Queen had just said.

"Well, but what about men? I mean real men?" Verity asked timidly. In Queen Alice's presence she had suddenly become unsure of herself.

"My dear," Lady Shy explained very, very kindly. "You are still an essentialist. Your daughter, Lady Verisimilitude, will explain it all to you. But now you must shush. The Queen is proposing to open Babel. The immigrants will be 'men' till they have been assimilated into the society of Babblers, at which time, at least some of them will definitely be accorded the status of women."

Shy sat down, and Alice glanced at Cinders and Little Red. "I know this is a question dear to your hearts." She leaned back a little and smiled graciously, "As a special concession to your seniority, an exception might be made. Your husbands, they may come in as women. Well, Cinders? Well, Red? What do you say?"

What could they say, poor things? Perhaps Alice was playing a cat and mouse game, perhaps she was being mildly sadistic; but if so, she seemed unaware.

Shy smiled at Cinders and rubbed it in. "Your loyalty to Babel must be beyond question, particularly if you hope to hold any office here."

Evidently dual citizenship was not being permitted. I admit I felt sorry for Cinders and Little Red. After all, what was there to choose between Wolf Man and Alice, or the noble Prince Charming and the good Queen of Babel? Rap Rap intervened.

"Your Majesty," she asked very carefully. "What is the point of letting men in?"

Alice sighed. She waved to Snow White, who got up at once and began to explain.

"Men are not men. That will be their designation when they first enter Babel. The Queen does not distinguish between men and rabbits. All are welcome, provided they've been screened and have scored a sufficient number of points on a suitable test. Babel will benefit by a trained workforce, which Babel has not had the expense of having educated."

"BUT MEN NEED EDUCATING!" Solly roared. She was beside herself and past caring what happened next.

What happened was that Lady Verisimilitude rushed in with a troop of bodyguards. Queen Alice remained unperturbed. "Let her speak," she commanded. It was the first and last speech in public that Solly was to make in Alice's reign.

"Men," Solly began, "men have diseased identities. No, I am not an essentialist. I mean that the identities on offer to men are diseased. I refer to men as they are at this moment, men with their mutated genes and memes, which have not mutated sufficiently. They are troglodytes still. They fight sabre toothed tigers when there are no sabre toothed tigers to fight. They worship power and find the victory of battle heroic. They batter one another, they batter piglets, parakeets and peonies, they batter women, they batter the whole of Crone Kronos' creation. This they perceive as the exercise of power; and when they exercise power, then they feel they truly live. They hate Babblers, they hate everyone whom they perceive as 'not them', but most of all they hate themselves. Even they understand that they are diseased, and yet they battle for their diseased being. To be cured is too onerous. Besides, power is an aphrodisiac. Power is a pheromone. Sometimes, they feel the intoxication of being in love with themselves. The smell of power makes their nostrils twitch. Why allow such creatures in?"

Queen Alice glanced at her. "Sit down, Solitude. You are a foolish woman to despise power. You do not understand it. Henceforth, be quiet. Should you disobey, I think you will find Snow White's silencers extremely effective." Snow White smiled deprecatingly.

Verity could see that things weren't going at all well for her friends. She had recovered a little from her awe of Queen Alice. Time, she thought, for one of her penetrating questions. Surely, the Queen would perceive that she, Verity, could penetrate to the very heart of a problem. "Your Majesty," she began respectfully, "what about the fact that some men might think that the rank of a man is higher than a woman's? I have often felt that had I been a man I might have served Babel more effectively."

"And so you shall be." This time there was no doubt about it: Alice could be nasty. "I confer upon you the attributes of manhood. Now leave. Your status precludes your attending this council."

Verity would have protested, but the look in Alice's eye made her reconsider. As she was led out, she pleaded for pardon and swore to serve as a woman eternally. Alice ignored her, and with a slight frown turned to Rose Green. "The problem of men? Will the memes hold? Or are they fracturing?"

Rose Green consulted her microcomputer. "There are stresses and strains, but they seem to be holding. We are monitoring the situation constantly."

Rap Rap rose. She had considered saying nothing. But what could she lose? Solitude had been silenced, Verity thrown out, and Cinders and Little Red barred from office. She might as well speak. "Your Majesty, opening the doors of Babel to men, that is to immigrants, is dangerous to Babel. Their memes will mix with our memes. There will be admixture and intermixture. Even pollution. And there is furthermore, the problem of an enormous influx of genes."

"All genes belong to the Queen," Lady Verisimilitude murmured, the bodyguards echoed her.

Queen Alice smiled. "Precisely. Rap Rap, you have forgotten the words of The Black Piglet. Metaphor in motion is metamorphosis. Whether Babel is composed of piglets or parakeets or immigrant men doesn't really matter. Under my guidance Babel will grow and thrive and live. My memes will win. That is how it will be. And generations to come will bless Queen Alice!"

"But Your Majesty, the memes will have been corrupted, mutated–"

"ENOUGH!" Alice was roaring. The bodyguards had seized Rap Rap and the others. Rose Green, Snow White and Lady Shy were keeping still.

"Your thinking is hopelessly counter-revolutionary. You are a pack of old women who have outlived your time. You are dismissed."

Alice swept away with a gaggle of courtiers trailing in her wake. As the old guard were led out, Verisimilitude whispered, "You were lucky to get off lightly."


Late that night as Solly and Rap Rap were sitting by the fire over a last cup of cocoa, I dropped in on them. I like fires. I was careful to remain unheard and unseen.

Solly was saying, "Are we really counter-revolutionary?"

"Retrogressive? Reactionary?" Rap Rap smiled, "I don't know. Best to go to sleep."

Solly attempted a small smile back. "In my dreams I'll keep Crone Kronos at bay,/But I'll wake again to Crone Kronos' day."

"Alice's day."

"Crone Kronos' day. It's not the same thing."

"Isn't it?"

They went to bed. The fire was still warm. I sat there wondering: In Rap Rap's universe would a cat be allowed to contribute to the memes?

A few days later I dropped in on Shy. It seemed only fair she be given a chance to show what she was like when she wasn't in public. I also had an ulterior motive. As Keeper of the Memes and Genes Shy had power. Surely disinterestedness did not require me to ignore my own species. I wanted to know what Shy thought of felines, of cats in particular, and of me especially. No harm in that. It was all information. As it happened I was pleasantly surprised. She was straightforward with me, even warm and friendly.

She had a well kept garden. I approved of that. And her house was clean and luxurious. It had central heating, air conditioning and soft surfaces. Little things matter. She offered me a choice of cushions and rugs, and asked if I cared for anything to eat. Don't misunderstand. I am morally certain she was not in any way attempting to bribe me. She was genuine. The fact is she liked cats. She was perfectly frank and open with me. She admitted right away she didn't care for women – too much backbiting, struggling and striving, and more often than not a sense of inferiority.

"What about men?" I asked her.

She liked them even less. She didn't bother to distinguish between Y chromosome creatures and immigrants to Babel. "Unbiddable and awkward. They don't fit in. Always in a mess, and always in need. More trouble than they're worth, but useful to the production and propagation of genes." She shrugged. "For the building of Babel probably necessary."

I then asked my main question. "And what about cats? What is the position of cats in Babel?"

"At present," she answered, "there are no cats who are citizens of Babel, and therefore, of course, the problem doesn't even exist." She smiled and added, "I personally like cats very much, and would be inclined to be in favour of letting them in – a selected few at the proper time."

I wasn't quite sure how to react to this. Why "a selected few"? What was wrong with cats? On the other hand, I was aware that I myself did not feel an undiscriminating love for my own species. I changed the subject and began to ask her about herself. What were her origins? Moderately middle class and reasonably honest. How had she found life in Babel? Difficult. But she had always been able to stand up for herself, and at present she was doing fine. My next question was intended to trip her. "Do you really care about the women who are less fortunate than you?" But she didn't find it necessary to pause and think.

"Yes," she replied. "Yes, I do. After all, I know what it's like."

I then asked her her real opinion of The Black Piglet.

She shrugged again. "A dilettante. An impractical idealist. Unfit to meddle in politics."

"And Queen Alice?" I thought I would get her this time. I waited to hear what she'd say.

Her tone was reproving. "Long live the Queen!"

"Yes." I replied. After all, I could hardly say, "No, let her not have a long life." That would not be disinterested. Besides, Alice and I go a long way back.

Shy smiled tactfully and changed the subject. "I've summoned all the poets of Babel tomorrow in order to regulate the matter of taxes. I would greatly appreciate your views on the matter."

I felt pleased. She had asked for my opinion, for a little sound advice. I would do my best to give it to her. "Well," I began, "the first thing to decide is who is a poet and who is not." To me this seemed like a sensible start. I hoped she was impressed.

She just nodded. "Oh, that's all right. Anyone who says she is a poet is a poet. Men, of course, will not be permitted to practise poetry, until they achieve the status of women. But other than that all self-styled poets will be duly recognised – and duly taxed."

I was confused, but didn't want to say so. I asked, "Taxed for what? For writing poetry? Or reading poetry?"

Lady Shy looked thoughtful. I felt relieved. I had asked a good question. "Originally, I had thought only of taxing the publication of poetry, but you're quite right. Ingestion of the stuff should be taxed as well. It would only be fair. Both activities constitute the use of memes, and all memes belong to the Queen."

"And it would increase your revenues as well," I pointed out. "What about the thinking of poetry? Should that be taxed?"

She shook her head regretfully. "No, the administrative costs would outweigh the benefits. At present, only the public use of memes is to be taxed."

"And public use–" I prompted her.

"Would consist of the communication of memes to five or more persons either individually or as an assembly gathered together for that purpose."

I had begun to understand the outlines of the scheme; and since I had some literary aspirations myself, there were aspects of the system that worried me. "How," I asked carefully, "how are the poets to pay their taxes?"

Lady Shy shrugged. "From their earnings, I imagine, when they engage in activities that are socially useful, and are not, as it were, merely indulging themselves."

"I see," I replied sagely. "So then no poets will be paid simply for producing the stuff?"

"Oh no," Shy answered. "For that they'll have to pay. Don't worry about it. It's unlikely to inhibit any of them."

I was feeling a little at a loss for words, but I needn't have worried. Shy went on. "You see, Queen Alice doesn't believe in curbing the freedom of the people. There will be no censorship. Anyone can say whatever they like, whenever they like and as often as they like. The memes are there for everyone to use."

"Provided they are willing and able to pay," I said sharply, then regretted it. My function, after all, was to be non-committal.

To my astonishment Lady Shy beamed. "That's the beauty of it! Without any censorship, the Queen with her revenues and her tax free status will, in fact, be able to control the cultivation of memes."

This time I really had nothing to say. I was imagining Babel thriving and growing in strict accordance with Queen Alice's plan. At last I murmured, "And will the Queen be hiring a chosen few, a select few, of these practitioners of poetry to write her own poems?"

Lady Shy frowned. I had gone too far. I was preparing to make a speedy exit, when she said,

"Yes," thoughtfully. Her face lighted up. "Yes. Why not? That would be a splendid plan."

She was effusive in her thanks. I had been most helpful. I was embarrassed. I was muddled. I found myself feeling pleased to be thanked. I felt awful. Narrators ought not to break down in public. I made use of my extraordinary ability to extricate myself from absolutely any situation. I vanished.

## 2

Narrators are people, creatures. Narration has a face, feelings, sensitivities. I needed a rest. When I felt strong again, I ventured into Babel. Much had happened. A riot was in progress. Little Red's daughter, Red Jr., had emigrated to Babel, and was by now the leader of the nobility. Oh yes, there was a nobility – mostly women, especially those with titles. Some of them were demanding control of their own eggs: all genes belong to the Queen except those of the upper nobility – something like that. Cinders had returned to Prince Charming, but came now and then to visit Babel, though her visits had grown more and more infrequent. Little Red, after the death of Erstwhile Wolf Man, had finally settled in Babel. She had lost her teeth, much of her bite, and on her hundredth birthday Alice had sent her a telegram. As for Rap Rap, she lived quietly in a small cottage near Little Red.

But I digress. The action was elsewhere. My own partiality ought not to dictate what I select. Babel had changed. Queen Alice's palace overtopped the Temple of Crone Kronos.There were some fine mansions with pleasant gardens, and there were acres of slums – well, not exactly slums, but a growth of mean, crooked houses with narrow streets and low standards of cleanliness. There were fields, farms, and country estates. Oh yes, and in genes and memes there was a roaring black market. The meme racket worked through downloading mostly. Disks – being physical objects – were more troublesome, though they were used to evade policing on the communications net. Genes – sperm mostly – were floated in, in plastic capsules. Poetic, I thought. I imagined the little capsules bobbing among the white maned waves, until I saw the beach for myself. Litter and mess. I was told that blackmarket genes were dangerous to use – no quality control. Also, Queen Alice had had the beach sprayed with a sterilising agent. There were riots of course, and demands for civil rights, particularly the right to bear children, and demands for a constitution and a reconstitution of the common wealth – all that stuff. Babel was bubbling and boiling, and striving and sprawling. Lots going on. Verity and Charity lived quietly together, and interfered with nothing. Verity's daughter, Lady Verisimilitude, had made it clear that her mother's insistence that she stood for the truth was now old hat: it was semblance that mattered. Verity had had trouble with this, but she had lost much of her old brashness and was content that her daughter was doing well. Charity looked the same as usual, a little sickly, but not disabled. Oh, and in the middle of all this, Solitude had died. It wasn't that it was a matter of no consequence. It was just that, at the time, her death hadn't attracted any notice. No doubt it was inscribed on the heart or the brain of Madame Mem. But if Mme Mem was no longer accessible to Solly, perhaps – it could be argued – it really didn't matter. I decided I would mention her death to Alice. After all, Solly's memes, with or without acknowledgement, had contributed a great deal to Babel.

Alice had altered. I chose a time when she was by herself. She was soaking in the bath tub. No *petit lever* or *grand lever* for her. Half an hour's privacy was a privilege, she told me, she had insisted on for herself. She didn't throw me out, just looked at me. "Oh, it's you," she said. "Yes." I said. She had grown more portly. She made no attempt to cover her breasts. But then why should she? I was only a cat. That reminded me, I ought to ask her: what did the Babblers do about sex? I know that for her species nakedness and sex seem to be closely connected. Perhaps not in Babel? For a while Alice and I said nothing at all. I scrutinised her – grey hair, flabby muscles, coarse skin, fur would have helped, but perhaps not while bathing – she didn't object. At last I asked her, "How are you?"

"Tired."

Another pause. I considered what to say. Well, what could I lose, I'd say what I thought. "You've made it to Queen, laid down the law, or, if you like, set out the ground rules. Are you content?"

She frowned. Then she muttered, "Too much untidinesss." It was as though she was too tired to explain.

I pressed her. "But as Crone Kronos' Rep. you have considerable power."

Alice shook her head impatiently. "I don't know that I believe in the old hag. Or if she exists, then I'm not convinced she's a beneficent goddess. I, Alice, believe in order, rules, rank and common sense. But now order has grown rank. The memes and genes grow this way and that. I don't know what causes it. Crone Kronos perhaps."

"Perhaps your notion of order was too static?"

Alice shook her head. "No, no. I took account of mobility. Laid down the paths and rules in accordance with which, those who were so minded, were allowed to move. But there's a lot of rule breaking, jumping and skipping, slipping and backsliding. There's all this nonsense about everybody making up their own rules or abiding only by the majority's rules. About how a cat is as good as a queen any day. That sort of thing."

It was clear that Alice couldn't understand why the Babblers were being so wrong headed. But Alice's inability to see why a cat was as good as a queen bothered me.

"Well," I challenged, "and isn't a cat as good as a queen?"

I thought I was in for an argument, but Alice didn't argue. She looked exasperated. "That isn't the point!"

"Well, then what is the point?" I retorted.

"The point is that the cat is not the queen. I am the Queen. If everyone did what they were supposed to do, there wouldn't be a problem."

I saw it was no use talking to Alice, so I said nothing. Alice persisted. "Come now, stop sulking. If you were the Queen, then of course I would say, 'A cat is as good as the Queen.' It would be common sense. You think I want everyone to obey me. It's not that simple, you know. I just want them to obey."

I admit I was getting more and more irritable. After all, I wasn't one of Queen Alice's subjects, and as a disinterested narrator I had the right to make judicious comments. I said, "It is that simple. You've come a long way, Alice, but the fact is you like power. You like control. You want to be in charge of everything that's happening. You want to be in charge of Babel, in short. But it hasn't worked, has it? Babel's in revolt and all your subjects are discontent."

I expected Alice to lose her temper; she only sighed. "Yes, I've come a long way, and sometimes I wish I hadn't." She began explaining things to me almost as though she was talking to herself. "Yes, I respect power. Power matters. That's why I laid down the channels so carefully, and made myself the source of power. Alice The Sun Queen. Alice The Successor to The Black Piglet. Alice the Embodiment of Crone Kronos in Babel. Well, why not? It was orderly. It was logical. But power is a meme, a whole cluster of memes. And so is desire. And the memes of power and desire have meshed. My subjects don't hate me. They reverence and adore me. They try to get closer and closer to me. They would, if they could, marry me. Or better still be me. In short I have a host of hungry Babblers to deal with who worship power."

"And so?" I encouraged.

"And so there is unrest and constant activity as they jostle and shove and plot and counterplot in order to get closer and closer to the source of power."

Watching her lying there in the bathtub looking flabby and tired I felt sorry for Alice. I said, "And so? You're lonely, Alice, is that the problem?"

She looked at me then. "Yes. Yes, I'm lonely. That is a problem. Now get out. I've finished my bath."

Well, I left her to it – her court and her courtiers. Later when I learned that Lady Shy had become more and more powerful, and then had suddenly fallen from power, I wasn't surprised. Who could Alice trust? The logic of her system made complete success for anyone but Alice – except on her deathbed – quite impossible. Poor, rich little Alice? And did I care? Sympathy was wasted. She would have claimed that in her reign Babel prospered. There was something about Alice I had begun to find tedious.

Surely somewhere in Babel the imagination stirred and broke through the painted stars and silvery moons gilding the hollow domes and minarets? I thought of Solitude in the desert, not mawkish and maudlin anymore, but severe and fine. I thought of Little Red, under that placid exterior, fierce and truculent and doing the work that came to hand, even if it meant clearing whole forests. I thought – I don't know. I thought of getting drunk that night in the taverns of Babel and getting the steam from Alice's bath out of my head. I thought that Babel was getting cheap and tawdry and that with one swipe of my paw I could knock it down. Would that be an exercise in the uses of power? Was I infected?

The tavern was called The Wolf's Gullet – appropriately enough, I suppose. Most of the customers seemed to be women. There were males about, functioning mostly as waiters, though one or two were seated at the tables. The noise level, the drug level and the smoke level was so high, I decided not to materialise after all. I hovered cautiously near a corner table. I noticed that every now and then a woman might pat a waiter's buttocks. Apparently, this was not considered offensive. The waiter would smile as though gratified, and carry on. Two of the waiters were female – immigrants, I realised, who had not as yet achieved the status of women. They too had their buttocks patted, though the tweaking and pinching happened less often to them.

Suddenly a dark haired immigrant (one of the females) convulsed and fell. The only possible cause was Rose Green's daughter, Green2, who was now fastidiously dusting her hands. I looked more closely. Red Jr. was remonstrating with her. They both stood up. Their hands were raised like those of boxers, but with this difference: their fingers were splayed, and their nails glinted and were sheathed in metal. Green2's sheaths were clearly brighter. It was this that seemed to make Red Jr. back off. She made her way to my corner table.

I materialised. "Claws?" I began conversationally, indicating her nail guards. I knew how effective claws could be. "Do all the women carry weapons?"

Red Jr. hadn't taken in at all what I had said. She was gawping at me. "By Croke!" she muttered. "A cat! I've had too much!" She straightened her shoulders, and decided to play along with whatever her fantasies had somehow invented. "Did you say something, Cat?"

"Yes," I replied. "Those claws? Does everyone have them? Do they function as equalisers?"

"Croke! No!" Red Jr. guffawed. "You mean the zappers. Their voltage is graded strictly by rank. That's why I had to back down just now. Green2 would have zapped me out – as she did that waiter."

"And the men," I said, indicating the waiters.

"Do they have zappers?"

Red Jr. stared. "Men with zappers? They wouldn't know what to do with them. They're afraid of zappers. And besides, it wouldn't be proper."

"So Green2 outranks you, does she?" I asked cautiously.

Red Jr. frowned. "Yeah. But I'm working on it."

"How?"

Red Jr. looked surprised. "By courting the Queen, of course. How else?"

"And?" I said encouragingly.

But Red Jr. cuffed me carelessly. "Curiosity will get you nowhere, cat. Go on, vamoose, disappear, scat."

I vamoosed, disappeared, scatted. And I was annoyed. Puerile idiots!

## 3

I looked in on those puerile idiots just once again. By then they were in the middle of a civil war each intent on outzapping the others. Green2 had inherited her mother's duchy and had done well out of the issue of licenses for memes and genes. She had made millions. Red Jr. too had done well. She controlled the trade in zappers. She told me that my question about whether zappers were equalisers had inspired her. I shrugged. Anything they did had ceased to outrage me. I had returned only because old Queen Alice had asked me to do so.

It's a fact. The message I had received was not a summons, it was a plea. *"Old friend, please come back. I have a favour to ask."* I admit I was flattered. So the Queen has something to ask the Cat. I think, on the whole. I really returned out of the goodness of my heart. Oh – since I'm being honest – curiosity also played a part. What did Alice want? I made conjectures. Perhaps she wanted me to succeed Lady Shy or whoever her current Chancellor was? Perhaps she wanted me to succeed her? The Cat to be the next Queen of Babel? Or perhaps she just needed a reliable confidante, a friend? What Alice actually wanted surprised me. It was something I would not have guessed at.

We met in the western turret of her palace. She had cordoned that off and had it debugged. The result was that it was rather dirty – no one ever entered except Queen Alice – but it was private. The good queen herself had aged. Her hair was snow white – she might, of course, have been wearing a wig, it was hard to tell. Her face was lined, though carefully made up. And she had lost a lot of weight; her bones showed through her papery skin. She cut a fine figure. She looked immensely dignified and very, very tired. So when she spoke gently and almost with humility, I was taken aback. She thanked me for coming.

"Oh, that's all right," I said. "It's good to see you, Alice. How are you?"

"I'm all right," she answered. "I'm preparing for my death." She didn't seem to be asking for sympathy, so I didn't offer any. Perhaps affairs of state were troubling her. Mentally, I pulled myself together in order to offer good advice.

"Is it the state of Babel that is troubling you?" I asked.

Queen Alice smiled a wry smile. "The state of Babel! The robber baronesses are killing one another. Some join one faction, then another. I let them do it. I watch calmly. Meanwhile, the women of property are demanding a voice for themselves. As for the worship of Crone Kronos, that has sub-divided into multiple sects. I've done what I could to contain Babel, but fungus grows on the inner walls, and ivy shatters the bricks of Babel."

"And this makes you sad?"

Alice shrugged. "The chess board has erupted. It wasn't what I intended, but I'm getting old. When I started out I gave them a good story: The Story of Good Queen Alice and Her Reign. Everything was so orderly. I set out the chess pieces on a black and white board. Crone Kronos and Co. in their proper places."

"You only had one set of pieces. You didn't allow for any kind of movement, any opposition."

"No, they were supposed to move. Like clockwork."

"But?"

"But Crone Kronos corrupted them. They had a motion all their own. They ran this way and that." Once again she shrugged.

It was obvious that whatever the problems of Babel, I wasn't going to be asked to solve them. What was it then? I waited. For a while she didn't say anything. She was struggling with herself. I might even have said – had I not known her better – that she was summoning courage. I'd never seen Alice look so awkward. It was disconcerting. At last she blurted out, "I've written a poem!"

"Yes?" I replied cautiously. "What's it about?"

"It's my elegy."

"Elegy for whom?"

"For myself."

She thrust a piece of paper at me. I was embarrassed. I read it carefully, trying to show by my demeanour that I was reading it purely as a poem, and that as a poem it had nothing to do with Alice's feelings, or her personal history, or, indeed, with any feelings I might have for Alice. I hoped she would take her cue from me. I finished reading it. It was quite good. I was surprised.

"It's quite good," I said judiciously to Alice.

She couldn't hide her relief. "I'm so glad you like it. I want you to publish it, to insert it into the network of memes."

I demurred. "The taxes . . . Why don't you publish it yourself? Your publications are tax free."

"No, no. I want you to publish my poem under your name. I'll pay the taxes. Please. That is why I asked you to come. That is the favour I want you to do me."

Now I was embarrassed. Did I want to put my name to Alice's poem? Was it good enough? Besides, would it be honest? And anyway, why was she so unwilling to claim it herself? Furthermore, if she no longer cared what happened to Babel, why was she so anxious to get it published? There was something very odd about Alice's request.

"Alice," I said. "It's not a bad poem. Why don't you put your own name to it?"

"Because," she replied, "my public image and my private poem are mismatched. I want this said, but I need someone else to say it for me."

"Your public image is your own creation."

"Yes. My public image: Good Queen Alice on the black and white checkerboard. But I am like the others, like the other chess pieces. They're not wooden characters, they sprout green leaves. They run here and there. They – how shall I say it? – they have their own feelings, think their own thoughts."

Good Queen Alice saying she was the same as everybody else! I suppressed a smile. "But why," I asked her, "why do you want to publish the poem?"

Alice frowned. She said slowly. "I'm dying. What else is there? I too am a person, not just a queen. I too was alive. What else is there of what it felt like?"

Well, it was awkward. She was breaking the rules. She was being personal, directly and definitely personal. In the end I muttered, "All right, I'll do what I can. I'll think about it."

I took away the poem. I read and re-read it. I even revised it, made one or two changes. And then, a few days after I had spoken with her, Queen Alice died. Did I publish the poem? Well, yes, I did. I inserted it into the memes of Babel (and I used the opportunity to insert my own narrative). I did not put my name to the poem. It was hers, after all – "All memes belong to the Queen". And I gave it a title, which made its subject matter, if not its authorship, quite plain.

### QUEEN ALICE'S ELEGY

Time was my true love, time my grief.
    A handsome young man? No, he changed,
        changed for his pleasure to a white rabbit.
Time was my solace, time my relief.
    A kindly grandam? An absent mum?
        Returned to comfort me, feed
            and fend for me, to the end
    I might grow to the height of her dreams?
        No, no such thing. A witch or a hag,
            smiling, oh so maliciously.
    "Look in the mirror," she says.
    "You, Sweet Alice, you are me."
Time was my friend, my good companion.
Time was the piglet, the little black dog,
    racing through the scrub, at my heel.
Time was my fellow who overtook me.
    "Come," says The Piglet, "come with me."
Time was my court, my consort,
    Time my courtiers. They did
        what they could. They flattered me.
Time was my city, time my realm.
Time was Babel; Time its crooked geometry.
    Time was the tower that stood up straight.
    Time was the tower that crumbled.
    Time was the tower that outlived me.

Queen Alice dies,
consigns herself –
    oh, no, no –
        not to her friends,
        not to her mother,
        not to the lover
        (who might have been),
        not to the fire,
        air or water,
          not to the sand.
She consigns herself to memory.
Let the three old women carry her away.
Alice is joined with her enemy.


I felt sad after Queen Alice's death.

She had arranged the obsequies with the same care as she had her coronation: a little pomp, a little splendour, and some pageantry. There were three pall bearers, Rose Green, Snow White and Lady Shy, creaking old women, quite suited to the part. I wondered for a moment whether they'd be able to carry poor Alice, but Alice was light. Alice had withered and shrunk, was a wisp of herself as she passed through the scenery. Poor old Alice. I find I liked her. Her passions were always impeccably orderly.

And where is Alice now? Romping about in the Elysian Fields, chatting with Crone Kronos and making the best of a bad job? She could always adapt, always adjust – maybe.

I didn't stay long in Babel after Alice's death. All my old friends were gone or turned into legends. Rap Rap, Cinders and even Little Red were revered for the greatness of their hearts, their rectitude and wisdom. The Black Piglet had become an icon. And Solitude? No, she was not entirely forgotten. She had been canonised and was now known as Saint Solicitude, The Black Piglet's faithful disciple. Fair enough, I suppose. No secret, creditable or otherwise, ever remains completely hidden. Any secret, every secret, once had a life. The life leaks out.

In Babel, Alice's death generated a period of hey-day and holiday and also of bloodshed, all mixed together. Some cried "Freedom!", some feared for their lives, and some did both. Red Jr., who ran the strongest of the baronies, looked set to win the civil wars. I think later Red declared a republic in Babel. I don't know. I left Babel. Alice was one thing, but I don't really care for children and childishness: the roistering and rowdiness, the "me! me! me!", the young blood and fresh blood, the angst and the ignorance. I don't like watching women – grown women – tearing each other to pieces, committing stupidities, committing crimes, and all in the name of Crone Kronos, The Black Piglet, Queen Alice, The Good of Babel, The Glory of Womanhood,

The Beauty of Nationhood, The Gorgeousness of Peoplehood . . . Babel was no longer distinguishable from any other city. Perhaps it would disintegrate, perhaps not. Who cared?

# V

## MAD MED

*Oh Baby, Baby, Baby,*
*Why have you set this in motion?*
*Because I wanted–*
*Pretty toys*
*for girls and boys.*
*Because I wanted to play.*

Mad Medusa sits like a baby among the building blocks. This is her inheritance, this the treasure, left her by her sisters. Mad Medusa, raped by Poseidon, betrayed by Athene, slaughtered with the aid of her own powers, surely the stupidest woman on earth, examines what is left. A is for Apple? Appeasement? Apathy? Apartheid? Armageddon? Art? Suddenly like a slow and giant baby's her face lights up. A is for Arrogance. She scrabbles among the bits for all the memes that might go with A, and stumbles upon B. She seizes that at once. It will fit into A in a dozen different ways. She moves her head in a slow scan. Her blue eyes are placid receptors. B is easy. B is for Babble, Babblers, Bombast, Bomb Blast, Bickering,

Battle, Bang Bang. For the time being she puts it down. There is more, of course. When she has fitted it all together, all she has to do is find a power source. But perhaps there is more than one way to put it together? And she has to find the right way? If all the king's men and all the king's horses had succeeded, they might have ended up with a different egg, a not-egg, an altogether different object, a multiplicity of objects, a fertile egg. A whole chicken farm? The possibilities . . . Medusa sighs and begins to suck one of the building blocks. A is for Amour, A is for Ardour, A is for Alice, the beautiful and passionate Queen of Babel. And B is for the Black Prince. B and P are both for the Black Prince. Which should she choose? A or B? She feels like being in love with both of them.

> *B is for Boyo. Beneath the brazen*
> *alphabet are a billion blowing roses,*
> *bursting and blowsy with life and love.*

Have roses bloomed in Babel? They must have done. Otherwise she would not have thought of them.

Medusa decides she will build Babel in the air. She has the bits and pieces. She will choose only the pretty bits. M is for Medusa. The parallel posts and the great V can be the main frame. It will rise high in the air. And its foundations? She is sitting in mud. Mother Earth will uphold her? Castles in the air require grounding. Is Crone Kronos the same as

Mme Earth? *No, Crone Kronos is greater. Crone Kronos has galactic connections.* M looks down at the rubble around her. She isn't a fool. It's Solly's memes telling her that. Sol, Solly, The Solipsistic One . . . She wonders what Solly was really like. Whether it's worthwhile making up a poem about Sol? She would rather think about Crone Kronos. Crone Kronos must be a great goddess with a crown of stars. No, with galaxies flaming from her hair. She thinks about her own writhing hair. Perhaps she, Med, is a type of Crone Kronos. But Mme Earth is more comforting. Perhaps Mme Earth is Crone Kronos' maternal aspect. *Who comforts you, Medusa? Who croons, Oh Baby, Baby, Baby, what are you doing now?* Lover Boy. But she can't find Lover Boy in the memes of Babel. And anyway Lover Boy wants her to be Mother. It's all inverted. Red Jr. is Lover Boy. And she can hear Red crooning, crooning and yearning, *Oh Baby Baby* . . . singing for his supper, singing for his mother. But she's allowed to bring her own feelings, bring her own longings to the memes of Babel.

Mad Med stands on her two feet and examines her situation. What can she work with? What has she got. 1. There's Mad Mem *[What?]* – herself and her experience. 2. There are the memes of Babel. And 3. There's mud. She would like someone to be nice to her. She once had a sister, two sisters, but one was kind. C is for Charity. She scrabbles through

143

the memes. Who does Charity connect with? Was The Piglet kind? Was Queen Alice? Whom did Charity love? Everyone? Medusa snorts. Everyone and no one. Sweet Sister Charity is hardly known to have spoken a word. But if Charity and Verity had spoken together, then what would they have said? What would they have said to her, Medusa? *Poor Baby*, they'd have said, *Poor Baby, Baby, Baby* . . . It's not quite what she wants.

Med says, "Sweet Sister Charity, why have you kept silent for so long?"

And Charity pats her on the cheek and strokes her wriggling, writhing hair, and replies, "There was work to do, and–"

"And what?"

"You wouldn't have listened."

"I'm listening now."

"All things shall be well–"

"Clichés!"

"Love thy sister?"

"More clichés!"

"Dear Mad Med, clichés is all I've got."

Obscurely, Med feels pacified. Charity might be her older sister, but she doesn't know everything. She, Med, should help Charity. She puts a protective arm around Charity's shoulder. "Listen, Big Sister. You do the feeling and I'll do the talking. Maybe then we'll arrive at something useful."

Charity sighs, and Med is indignant.

"Why are you sighing?"

"I've heard it before."

"What?"

"What you said. That you'll do the talking and I do the feeling."

"From whom?"

"From Verity."

"Oh." This stalls Med. What is her relation to Verity? And does it bear thinking about?

She begins to make up a story about them. "Once upon a time there were two sisters, Maddy and Very. Very was the good sister. She did what she was supposed to do, said what she was supposed to say, and on the whole, she caused very little trouble. Maddy, on the other hand, was very, very difficult. She did what she liked, played with the boys, and went about saying she didn't give a damn. In the end she ran away, and everybody said, 'Good riddance'. Very carried on, carefully not bugging anyone. Then, one day, they got a fax: Maddy was in trouble, could Very help. And Very? Very didn't do anything at all. Just looked sorrowful – like everyone else."

To this Verity replies: "Oh for The Crone's sake, Med, someone has got to keep their balance. Someone has got to give this family some credibility. Someone has got to say what's what. Someone has got to agree with everyone!"

"Why?"

"In order to get on."

"Get where?"

"On."

"Huh!" Mem gets cross and is about to kick the rubble when she thinks of something. "Very," she says sweetly, "Very, you are past mistress of saying exactly what people want, then say what I want. Speak sweetly to me. Tell me I'm wonderful!"

Verity pats her cheek though she can't quite bring herself to stroke her hair. "Dear, little Med, sweet, little Med, you are wonderful!"

Med smiles. If Verity looked closely, she would see a malicious glitter in Medusa's eyes, but Verity doesn't look closely, she never looks closely, and anyway it isn't useful. Med says, "Oh Verity, you say the nicest things. Tell me the thousand and one ways in which I am wonderful."

Med expects Verity to get cross, but Verity doesn't. Instead, she sorts through the memes and picks out a handful of poems in praise of Queen Alice. She gives them to Med. "Just change the name and play them over and over and over in your head."

Med frowns. "But that's solipsistic. I need you. I need someone else to tell me I'm wonderful!"

Med kicks the rubble. Women are no use! Men. Now men might be different. At least before they do her in – again and again and again – men can be a load of fun. And they're beautiful! Her Boyo, her Yobo . . . Med looks dewy eyed. But there are no men in Babel. Yes, there are. No, there aren't. Not real men. She petitions Crone Kronos.

"This place is effete. This place is effeminate. If you want me to play with the memes of Babel, I must have my Man."

"Who is your man?"

"He is my Boyo. He is my yobo. He is a prince, reduced by Circe to the status of a Piglet."

"Indeed." The Black Piglet presents herself to Med in the form of The Black Prince. She is very handsome, though still a little effeminate.

Med decides that The Black Prince will do. "I want you to love me, to chase me, to overpower me," she informs him.

"These things are not synonymous," The Prince says gravely. "An imbalance of power . . ."

"I too have power," Medusa says proudly. "After all, you will be doing what I want you to do."

"I see," replies The Piglet. He tries to hide the distaste he feels for Medusa. *Bossy, little bitch!* The words come unbidden into his head. He doesn't like Medusa much. He likes himself even less. If only he could teach her a lesson. Medusa is tall. But he is taller. A simple fact. She will have to look up to him. He will make use of that. He will triumph. And Medusa's blinding blue eyes blaze, and The Piglet shrivels. A pity. But it wasn't always like that. Medusa remembers her own battered and bloody head. Sex and passion. Sex and blood. That's how it has to be. There has to be adrenalin before anything can happen. She kicks The Black Prince. "Come on. Fight and get up. Come on. Fight and get up. Prove you're a man."

From the mud The Black Prince says, "I'm not a man."

And Medusa says, "Come on. Fight and get up. Come on. Fight and get up. Come on. Fight and get up."

The Black Prince squirms. Medusa squirms. After a while they both squirm. They squirm in the mud. They both hope that Mme Earth doesn't mind too much.

When they finish, they both feel better. They want to get married. They want someone's blessings. They want their friends. They want Babel to rise again. But this time it's got to be better. This time different. This time they will make Charity Queen. This time Crone Kronos will be an aspect of Charity, not Charity of Kronos – that is if she ever was. This time they'll plant seeds, not memes, and the seeds will be intelligent and not make mistakes, and Babel will be a flowering desert, and – and – they both think that they are wonderful. After a while they do it again. And it is wonderful.

It occurs to Medusa that at times at least she ought to be awkward and austere like Solitude, not spend all her time on pretty toy boys. D is for Dunce, Donkey, Desert. No, that won't do. S is for

Solitude. She ought to ask an intelligent question, e.g. what did Solitude do in the desert? She masturbated. No, that's very wrong. She mustn't say that. Well, anyhow Solly wouldn't like it. Solly wouldn't inhabit the same space as that. It's – it's a breach of decorum. But what did Solitude do in the desert? She cajoled Death. She sang: *Oh Baby, Baby, Baby, / Your belly is full* – Nothing rhymes with "full" except "wool", and "sheep" rhymes with "sleep". *Oh Baby, Baby, / You should definitely go to sleep.* But Death isn't like that. Death is slow and sadistic. Death comes with a sword and rips you up. D is for Damsel and D is for Death. And S is for Sword and S is for Sex. But Sweet Sister Sol wouldn't have known that. It's no good. She has a feeling that Sweet Sister Sol doesn't/ wouldn't/ can't like her much. Wherein lies the problem? She, Sol would think that she, Med, wasn't – wasn't quite – wasn't? *Not poetic enough.*

Not? Not poetic enough! I'll show you. I'll show them! I'll build Babel six times over. And I'll build better than any of them.

"**Babel One:** When Rap Rap, Cinders and Little Red started to build Babel, they decided that the first thing they needed was a well, the second thing they needed was a garden and the third thing they needed was a wall. Accordingly, they found a well, made a garden, and built a wall. Then they took a rest, and sat about chatting around the well. After a while, Cinders said, 'What about the tower?' 'We're building it,' Rap Rap replied. 'Every word we say drops into the well like a small pebble. Soon the pebbles will rise above our heads.' In time, that was exactly what happened. And in time, when there was no longer any water left, Babel perished. But not in their lifetimes.

**Babel Two**"

"Hang on!" Cinders, Rap Rap and Little Red don't know whether to be amused or annoyed. They leave it to Little Red who has had experience with her own unvarnished young.

Med glares at them in triumph. "See! Out of a few elements Babel rose and Babel fell."

"Where did you find the well?" Little Red speaks quite gently.

Medusa crows. "I put it there. An omission of yours which I rectified."

Cinders, Rap Rap and Little Red don't say anything.

Medusa adds with kindly patronage, "Water imagery is a necessity." When they still say nothing she shrugs and mutters, "It was probably implicit."

Rap Rap, Cinders and Little Red just stand there. Medusa squats in the mud again. She looks up at them. "There's such a thing as progress, you know," she informs them cheekily.

At last Little Red speaks. "Medusa, we didn't just sit beside a well and chat endlessly. We built Babel."

Med: (*adopting the manner of a six year old*) "'swat I said."

Cinders decides to take a hand. "Medusa, how would you like it if we decided to make up stories about you?"

Med: (*still in the manner of a six year old*) "I can make up my own stories. Once upon a time there were three sisters." (*She sticks three memes in the sand.*) "See. C is for Charity. F is for Verity." (*In a different voice.*) "Truth is really a matter of faith. It's all that's left. And M is for Mad Hope. That's me. The voice of your future. If I was standing, I'd bow!"

Rap Rap: (*sharply*) "Pull yourself together, Medusa! You're not what we intended."

Med: (*nastily*) "No? But I am what you were. Luckily for me, only in part. It's not the Crone who would like to create in her own image. It's you. The Crone is an optimist. An exuberant experimentalist. She carries on."

Cinders: (*calmly*) "So you repudiate us?"

Med: (*reverting to her six year old self*) "You don't like me. It's not good for me not to be liked. You probably hate me!"

Little Red: "What makes you think we hate you, Medusa?"

Med: "Because I'm alive and you're dead!"

Having killed off her ancestors, such as they were, Medusa feels bereft. Now what? Perhaps she ought to re-make them? The thought occurs to her that perhaps she ought to re-make herself. But that is too hard. She has her bodily needs, after all: the softness of the squishy mud, the warmth of the sun on her breasts, the berries, the juicy berries, which have grown so unobtrusively while she wrestled with thought, and which dangle so close above her mouth. Perhaps she ought to kill herself? She could make up a story about how that came to pass. About how they built Babel, and about how Mad Medusa threw herself down from the very top of the tower. And when she hit bottom, then what happened? Then Crone Kronos picked her up. "*Poor Baby*," she said, "*Baby, Baby*," and rocked her in her arms.

Or she could leave the tower unbuilt and just have the garden, the wall and the well. Then she could throw herself into the well, drown in the deep, dark, doleful waters. Or she could pole-vault enterprisingly – over the wall, high into the sky, straight into the arms of Kronos herself. "Oh Daddy, Daddy," she would say. "I, me, we are so big and strong! This is heaven. Oh, let it last!"

Or she could finish telling her tale.

**Babel Two** As Rap Rap, Cinders and Little Red sat beside their well trying to decide what to do next, a man came along and laughed at them. "Babel wasn't built by the gossip of women. You have to be able to take the measure of mud. The building of

Babel requires brute strength." The women watched him flex his muscles. Convinced he was right, they sold themselves into slavery and let him get on with building Babel.

Medusa sighs. A story about how they didn't build Babel is also a story about the building of it. She props it in the mud next to her first story. After all, she too could build a wall, even a garden. Crone Kronos or Mme Earth – with or without galactic connections – could supply the well. If it came to that, she could dig it herself. This time she begins sorting through the memes more carefully. She even prepares seed beds for them. Some slip through her fingers and are trampled underfoot. Medusa shrugs. Well, why not? She can't keep an eye on every single meme. The more the merrier. Medusa frowns. Gardening requires controlled murder. Some memes shall grow and others shall not. Why not leave it to Crone Kronos of whom it could be said that she is truly disinterested. There would, of course, be murder and mayhem. But since Sister Death, follows the old crone and kills everything, there is presumably a massive justice in that. To be truly moral she must sit in the mud and do nothing. She need not squat, she need not sprawl. She could sit elegantly in the lotus position and listen to the sound of memes growing: massive earthworks, earth and stones being shoved aside, and fabulous artefacts rearing in the air. As easy as that! Culture does not require cultivation. Medusa sleeps. Meanwhile – though no Sleeping Beauty, but doing her best and this without any effort – meanwhile, flowers and weeds creep up on her, so that when she wakes up, Medusa feels that she has somehow been rewarded for being herself.

If, she, Medusa, built well, if she made a truly beautiful garden, then perhaps murder would be forgiven her? By whom? By the Queen perhaps? For the Queen she would paint roses. She would trim hedgerows. She would set out garden paths. She would arrange for her a fanfare of trumpets – done with daffodils, tulips, all in accordance with the Queen's fancy or fantasy, her whim or whimsy, or even, if required, her sense of propriety. She would find flamingoes, croquet balls, hedgepigs anything. Well, why not? It could all be made to fit. With enough skill, will, good will, nothing, nothing at all is not in good taste. And what might be the Queen's pleasure? Tarts. Jam tarts. And English tea. Mad Med serving tea to Alice? No, no, not bizarre. Well, not inappropriately bizarre. Suppose Alice were asked, "Listen, which is odder? Tea with a dormouse who's fast asleep? Or tea with Med, salty from the sea?" Such a question wouldn't faze Alice. She would drink her tea with queenly *savoir faire*, royal aplomb – whatever it takes. And if Med had got it wrong, then she, Alice, would make the gardeners – somehow, wherever Alice was, there would always be gardeners – would make those gardeners paint the roses! Very sweet and very bloody. O sweet and bloody Alice. O paramour, so primarily concerned with mud and death. Otherwise why thorns, why heraldry, why chivalry, or the burial of the dead?

"Don't be silly, Med. We only bury the dead, because they are dead. We don't kill them – in order to produce roses."

"Why not? We might as well recycle, re-invent . . ." Med examines Alice, who looks more like the girl in the Teniel illustrations than like Queen Victoria. She also looks willing to squat beside Med and play in the mud. An egalitarian Alice! Med is impressed, but soon discovers that Alice is bossy. She informs Med that she, Alice, is not particularly interested in mud or blood.

"Well, what are you interested in then?" inquires Med.

"Good government."

"Why?"

"Because it's safer. I like things to be under proper control."

"You must be power mad."

"I'm quite timid really – like the white rabbit," Alice replies, "and a little nervous."

"Why?"

"Because people are incalculable."

"Well, I personally do not want to be a rabbit. Why can't we be flowers? Two delicate flowers blooming in the fields?"

"What will we do when Crone Kronos strolls by? Bend, but not break? Duck if we can? Or bury our heads – as though we were bashful – in the oozy mud?"

Med: Yes, why not? Anything that works.

Alice: How do we know that this mud will suit us?

Med: If it doesn't, we'll fertilise it to make it suit us!

Alice: So we're plants who make their own fertilisers. I thought plants ate chemicals?

Med: Yes, well, they do both.

Alice: Are their memes chemicals?

Med: Ye-es. They eat memes and they secrete memes. They swallow culture and secrete culture, the artefacts of culture. There's memes and there's mud. I've just understood! The memes are mud! The mud is composed of memes! And memes in motion constitute space and time. Don't you see? Because the memes whizz about and form combinations they have to make a space to do this in. The motion of the memes generates the Crone! Like Aphrodite rising from a foaming sea! Like you and me sprouting from seething mud!

Alice: What do you mean?

Med: I mean that Crone Kronos is composed of mud. You are composed of mud. I am composed of mud. Mud is composed of mud! It's glorious!

Alice (doubtfully): Why is it glorious?

Med: Because I've just understood it. Come on, dig your toes in. This soil is chock full of memes. It is memes. We could flourish here.

Alice: What if the memes poison us?

Med: They might. But after all, they are our memes.

Alice and Med stand in the mud and are aware that time is being generated. But it's very slow.

Alice: Med, if anyone saw us, would they say, "Oh look, two charming little girls standing in the sand like delicate flowers. Aren't they lovely!"?

Med (gruffly): I'm not a flower. I mean to survive – just like a weed.

Alice: Yes, me too. I mean to survive.

More awareness. More generation. Alice and Med wriggle their toes. Then Alice says, "Let's run about. Let's do something."

"Why?"

"Because I'm bored."

"Okay. Instead of just letting the memes generate, let's exercise a little control."

Med and Alice step out of the mud. They make a garden, build a wall, and dig a well. They have babies, they regulate gender. They have sex. They change sex. They squabble over words. They fight furiously. They try very hard. They build Babel. They do what they like. They do what they can. They do not do very well. They try again.

Meanwhile, in the short term, in the interlude so to speak, the garden is a place of idyllic splendour. Everyone exists, everyone persists. Everyone and everything get on with each other. What do they do in this garden? They sing songs. They drink wine. They make love all day long. It is a Pleasure Garden. What else do they do in this garden? They think thoughts. They meditate. They levitate. They delight in the beauty and complexity of the universe, for which they do not think they are primarily responsible. It is a Garden of the Intellect. Nothing else? Don't they garden in this garden? Oh yes, they garden like anything. They tend and nurture and take extremely good care of the plants and animals and of themselves. In this garden no murder is necessary. This fact, namely that they do not have to commit murder, makes them almost delirious. It is a Moral Garden. But don't they socialise? How do they get on? Do they have parties? Governments? Groups? Gangs? Mobs? Who's in charge of law and order? And who pays the taxes? Are they capitalists or communists? Or socialists perhaps? What are their politics? What about economics? No, no, no, it's not like that. Everyone has the same power. You mean it's a democracy – everyone has a vote? No, everyone has a loud voice and in everyone's finger-tips there's the power to stun. And so? And so there are very few arguments because they'd deafen every-one. And the stunners? There's not much point in being stunned, is there? Well, but don't people gang up? No, the people don't gang up, the plants don't gang up, the animals don't gang up. Everyone's an individual. What about over-crowding? Space-time expands to accommodate all the life available. I told you, they have Crone Kronos thoroughly and properly under control. Well, what about the sick animal, the diseased plant, the frightened human? What about the uncontrolled exercise of power in order to ward off uncontrollable fear? No problem. Every creature is put into therapy the moment it's born – just in case. Well, all right, but what about basic necessity? What about food? What do they do when in order to generate a bit more time they are forced – reluctantly – to eat each other? No, they're not. They all live on pellets scientifically prepared from first grade mud, which are further fortified with memes, and vitamins and sundry chemicals. Well, all right, but what about Death? Don't be silly in the earthly paradise they generate time. As long as they're healthy in mind and body, they can go on forever or for as long as they like, change with the seasons, spiral and swerve, metamorphose – it's one long festival. What if they're not? Not what? Healthy. Why wouldn't they be? What if they choose? Choose what? Not to be healthy. Why would they choose that? Well, but suppose they did. Then the garden goes to seed, fruit rots, under-growth grows and everyone has to spend a lot of time trying to recreate the garden, but it doesn't work. Why not? Because now they're spending time, earning time, hoarding time, all that sort of thing. They're not making it. What difference does that make? Now they're beggars. Don't you understand, it's a difference in status. But beggars can make good. Yes, they dream about that. Some succeed. Some beggars become richer than others. Why do you call them beggars? It's very cynical. No, no, they are only beggars in comparison with what they once were. In another of their dreams, they see themselves as tragical.

Who wrote that? Who said that? Med and Alice know perfectly well that they themselves are not entirely responsible. They also know that malice cannot be apportioned easily. They are young women now, bright and polite, like postgraduate students, and chock full of memes. They know that the forecast isn't based on the entrails of a goat, or on their own entrails, but on the logic of dreams. They feel, however, that a less cynical and tragical outlook might produce a less cynical and tragical outcome, and that they are the women for the undertaking.

Alice frowns. "The problem is, Med, that the earthly paradise is contingent on Crone Kronos remaining beneficent. Once things go wrong the question arises: which is wiser – to kill Crone

Kronos or to placate her?"

Medusa frowns "Is she killable?"

"'Crone Kronos is dead. Crone Kronos is dead.' All we have to do is repeat that over and over."

For a time no one speaks. Alice is gazing at her own future. Eventually she says, "But if Crone Kronos is dead, then it falls on my shoulders – and on yours, of course–" she adds kindly – "to keep things going. Let me quote from the text: *They delight in the beauty and complexity of the universe, for which they do not think they are primarily responsible.*"

Med murmurs quietly. "In paradise we were children, and Crone Kronos was Mummy. The milk from her breasts seemed inexhaustible."

"Well, we're grown ups now," Alice is inclined to be a trifle sententious. "We will have to take responsibility for who we are and what we become."

"What about other people?"

"What about them?" Alice shrugs, and gets on with the job of being someone.

She decides that this time around she's going to be the Power behind the Throne. She acquires control over an insurance company, a pension fund, two supermarket chains, several media interests, an investment corporation and a brokerage service for information. Other ventures include education, medicine, and communications, together with water, electricity, gambling and guns. She can tax her subjects

– through the insurance companies and pension funds – without their chafing under the burden. She sees herself as a kindly parent: one, who allows her children to feel independent. During Alice's reign there are a number of wars, famines and massacres, but Alice makes sure that profits stay up; and towards the end of her life she establishes a number of charitable foundations.

Med, on the other hand, is rather less successful. She sets herself the task of learning and understanding what she can. She writes poetry, studies history, toys with technology, and, in her own way, has a great deal of fun. She does not, unfortunately, make her fortune. And since most of her work remains unpublished, any influence she might have had remains uncertain. Towards the end of her life when she is truly desperate, she applies for a grant to an Alice foundation. With, what Med feels at the time is astonishing generosity, the grant she has applied for is given to her. This enables Med to continue living as she has hitherto done, to the end that her life, which has been in its way a colossal failure, remains unremarkable.

Having worked out their lifelines to their own satisfaction, Alice and Med sit on the edge of the known universe and dangle their legs over the abyss. They ponder further.

"So," Med begins. "That was life on earth. But in

that scenario is Crone Kronos dead? Did we succeed in killing her?"

Alice's reply is slow and reluctant. "I don't think so. I think what we did was lobotomise her."

"Explain."

"The Crone was like a milch cow in an automated dairy. She produced milk and we consumed it daily, in quantified measures."

Medusa frowns and glances at her friend. "Not – how shall I say it? – not much fun."

Alice nods. "No."

Med tries again. "Well, we've had life on earth, and the earthly paradise. What about eternity? What happens then?"

Alice shrugs. "I don't know. This, I suppose. You and I perched in outer space. Galaxies whirling like Catherine wheels. And Rap Rap and the others whispering tunefully in our untuned ears."

For a moment both Med and Alice feel that they'd prefer the certainties of life on earth. Still, onward and upward, forward and backward, into the breach, words and thoughts can fill anything, even eternity. Med feels muddled. Stars could drown one. Space and time could fill one's lungs – like water. It is with immense difficulty that Med gasps, "What about Crone Kronos? In eternity what happens to her?"

"In eternity," Crone Kronos sings – does she sing

or does she laugh? – "in eternity I become the flute-playing god, the foam-flecked goddess. I become The Seducer, transcending, always just transcending, the imagination."

Med's head tilts back as her nostrils fill with Crone Kronos' breath. This is too delicate, too difficult, too intimate. Essence of Time. What a fortune such a perfume would make on earth! Crone Kronos' body is the air enveloping her. Ah, yes, yes! Crone Kronos is a drug. The most potent drug ever invented, found, discovered. A drug to which her brain is susceptible. To which her mouth, her tongue, to which she is susceptible. Crone Kronos in the wink of an eye. Crone Kronos the berry upon her tongue, which her teeth break open. To lie in the mud, while Crone Kronos sings to all her senses, while Crone Kronos coaxes and cajoles her, while he runs about and plays with her and teases her, that is something, that is much. Where are the others? Here, stroking her hair, caressing her. Eternity is probably not very proper, but it is blissful.

Medusa rises. A bit of eternity goes a long way. Meanwhile, there's work to be done: tinkering and touching up, bolstering Babel, or whatever.

# VI

## PATCHED PIECE

### I    TWO SISTERS

i)

Rose Green said, "If you were boss of the whole planet, if you were queen, if you had everything and I had nothing, then I would be nothing. That doesn't seem fair to me."

Snow White replied, "Don't be silly. All we have is our own bodies, the breath we breathe, and Crone Kronos' permission to continue to live. The rest just flows and floats with us, is contiguous to us. We don't own it."

"Nevertheless," Rose Green insisted, "I would very much like the title deeds."

"Have them," shrugged Snow White.

Rose Green demurred, "No, you have them too."

"Why?"

"Because then you'll have a stake in the legality of it."

"That is very sisterly of you," Snow White murmured.

Unabashed, Rose Green replied, "Yes, it is. In its own way it really is."

ii)
Having divided up the world to their own satisfaction the sisters have time to amuse themselves.

"What shall we play at?" Rose Green enquires.

"Piglets and Parakeets."

"How?"

"I'll show you. You run around very fast, and try to touch everything you possibly can. Everything you touch turns into a Piglet."

"And everything you touch becomes a Parakeet. Then we count them and see who wins?"

"No. I turn into a Parakeet, perch overhead and watch you running."

"You think you're superior, don't you?"

"Yes."

"Well, I'll turn into something that flies higher and faster, faster and higher, than any Parakeet!"

"I'll turn into an eagle!"

"I'll turn into a spaceship!"

They shoot into the sky. They become mere specks. They have ambitions and aspirations. Soon they can't breathe. No one has flown so high before. There's no one to see them, no one to appreciate the skill, the verve, the audacity of it. They turn to each other on the currents of air and bank expertly.

## II   BEING LONELY

She cradles the moon,
     chats with tadpoles,
and when she's tired,
     falls asleep quietly,
under the ledge
     of a neighbouring rock.
*"The birds have feathered
     their own nests,"*
          the sly wind croons.
She cuffs the wind,
     makes him her pillow.
She mumbles in her sleep,
     "Why should they not?"

## III   RABBIT OR WHATEVER

i)
In the forest
        both fawn and I
                lose ourselves.
We like one another
and are excellent friends.
            But afterwards
it's not so pleasant.
The cat is a cat.
        Cats eat mice,
        sneer at girls.
        Rabbits run away.
        Girls scare fawns.
        Fawns run away.
        Caterpillars growl,
        are extremely rude.
        Queens kill people.
        Maids scrub floors.
I want to go back to that lost forest
        where metaphors mix,
        rub shoulders with each other,
and everything turns to everything else.

ii)

Then it happened. In the course of her wanderings Alice came across a motherly person. It was Little Red, thoroughly grown up, thoroughly tired, but with a little bit of kindness left still. The white rabbit scuttled up to her, the gryphon, the gorgon, Alice herself, even the cat who liked her independence, they all of them scuttled, the king and queen, even the gardeners. Little Red sighed. She gave them each a kiss and a hug and stroked their heads.

## IV   SUCCESS SECTION

*But the roar of the guns still outroars.*

Three hags sit about a fire in the middle of a forest. But why these dress ups, set ups, hang ups? The fire is necessary: it keeps them warm. The forest is there: Little Red has saved it for herself. It has a lodge, and the other two are Little Red's guests. As for being old, that is not something Cinders, Rap Rap or Little Red can help. "We meant well," Cinders offers. "What difference does that make?" Rap Rap mutters. Cinders shrugs, "We'd have done a lot worse if we had meant ill." Rap Rap glares. Little Red sighs. In a way none of it matters, but she is the hostess.

Sadly she says, "Sweet Sister Success doesn't rise ever higher into graceful old age. Crone Kronos comes along. What we had worked for somehow mutates."

Rap Rap says, "Then Mad Med comes along: boisterous, ebullient. 'I am of supreme importance. Where is my inheritance? Where my sustenance?' She snatches our bread. She stomps about. She makes a mockery of all we've made."

Cinders sighs, "She is what we've made."

Three old women stare into the fire. There's a crashing in the darkness. Snow White and Rose Green. They want to share the fire. Alice and the fawn. Mad Med & Co. The three old witches are tempted to set the forest alight. Rap Rap brandishes a burning brand, but only in her mind's eye, only for a moment. They break the circle. They shift a little, and make space.

## V   A REMOTE PALM

Meanwhile, Charity and Verity under the shade of a remote palm watch Sister Solitude surfing the sand dunes.

"Do you suppose she's enjoying herself?"

Verity stares. Not like her sister to question anything, or to smile a smile that is less than earnest.

"Who knows," she grins, and watches the arabesques.

A speck on the horizon. Lady Shy riding the wind, sowing the whirlwind. The sand hisses by like particles of metal. What fuels her fury? She does. Lady S and Sister S carving up sand dunes, demarcating boundaries, sawing up space. Soulmates possibly, even sisters.

"Why make the sand sizzle past fast?"

"Why not examine one grain of sand?"

Shy and Solly hop off their sand bikes. They cultivate their garden. They grow memes that are fat and round.

And beneath yet another palm, The Piglet and her Death curled up together: they sleep and dream and chat in the shade.

"It's not just that we eat each other," The Black Piglet mutters. "The future eats the past." She snuggles up closer. "That's how it is. Can't be helped."

"So you forgive Mad Med?"

"I haven't even met her."

"Don't quibble."

"All right, I forgive Mad Med."

"Does she forgive you?"

"Mad Med is ignorant. She doesn't even know she has been forgiven, nor does she understand what she has to forgive – yet."

## VI  ADVENT

In the difficult distance a dome gleams
        on the sea's horizon.
Babel rising? The sun or the moon?
But no floating city drifts to their shore.
        No metaphors now.
It's Mad Med cradled in an eggshell.
        Awake or sleeping?
 No hi fi blasts the air.
 No seaweeds writhe in her hair.
And the waves are so gentle that the sisters
        expect the sound of a harp,
                well, some heavenly music.
 "Mad Med, welcome ashore!
        Mad Med, you bloody–!"
No, no such sounds, only the sound
                of waves lapping.
Since the sisters keep their grief to themselves,
and forgive Medusa her ignorant beauty.

VII

*"Compleynt, compleynt, I hearde upon a day"*
[*Canto 30, Ezra Pound*]

So it's my birthday and all these godmothers come
to wish me ill or well – depends on the time of day,
and how they're feeling. Little Red brings a tinder
box. I'm supposed to be grateful? Rap Rap has
brains, Rap Rap's a scholar. She brings a computer. I
play games. Rap Rap scowls. She had wanted me to
acquire wisdom and learning. I am learning. And
Cinders is superior. Cinders produces a party dress.
That's more like it. It dazzles and glitters. Sequins
and silk. When I open it up, it disintegrates. Can't
even sell it. Well, they meant well. I'm supposed to
smile. I smile a bit. Then I cry. I do what the young
are supposed to do. They smile. They scowl. Two
can play the game. Why not? Solitude brings me her
solitude. The Black Piglet? Tells more stories. Oh
well. They bring what they can. I take what I can. I
rip out the sequins. Someday I'll make a brand new
dress – something that fits.

I DON'T forgive them my impoverishment.

## VIII   AND THE CAT SAID

Am I to be excluded from this precious paradise? One little contretemps would blow it sky high. Is it because I'm only a cat? Or is it because I'm more than a cat? I have a personality? I make an impression. Despite text, context and the grounding of grammar, perhaps I come across as a beer swilling, whisky drinking, journalistic cat, even a poet? What is the problem? Am I a TOM CAT? Who Defines, Defies, Deifies me? O Sisters, you are my sisters, are you not? Yes, yes, I know. The problem is I am not your sister. Am I your brother? If I am willing to be your sister, then will you have me? Rehabilitate me? Resurrect me? Correct and cajole till I fit in? O Sisters, in this season of happy humanity is there no room for an ordinary cat, a spat upon, scarred and scabrous cat? A bad cat? You say you will stroke me? Despite my ill humour? Well, go ahead. I dare you. See what you get. It worked you say. Here I am, purring and basking in the light of your eyes. Sometimes it works. But ladies, do not forget the logic of grammar. Sometimes it works, and sometimes–. Yes, yes. Extremely pleasant. Stroke my tummy. Just a little higher. You think you have tamed me. Don't you believe it. You say you do not. Good. Good. That is excellent. Just a little lower. Just a little harder. Making me happy – that's what it's about. Is it not? You've accomplished something.

You have pleased a cat. Try boys next. Girls. Try women. Try men. You think I'm scoffing? No, no, can't you see? I have already declared my interest.

Alice to Cat: "Come on, Cat, grow up."

"Why?"

"Because otherwise the sun will stop shining. The stars will wink out. And we'll all die."

"We'll die anyway."

"Yes, but let Crone Kronos harvest us. It isn't necessary to kill ourselves. That isn't demanded."

## IX  BABEL SPROUTING

In paradise, is it permitted
that the sisters mourn? The desert
has sprouted willow trees; a deep,
a wide, a powerful river runs
on and on. The soil is full
of surprises. Someone is going
to live. Someone is going
to die. A determined few
are determined. They don't know
why. Should the sisters plough
the desert? Should their mood
be plangent and sad? What should
the sisters do? Is it seemly?
Should they be glad?

## X    A TEXT DOESN'T EXIST

The sisters pull themselves together. So much more pleasant to lie in the shade. Let the building of Babel regulate itself. But The Piglet wriggles from the arms of death. Solly and Shy – for a time at least – refrain from gardening. And Rap Rap, Cinders and Little Red present themselves.

"Okay," they say to Capital C Cat. "Be a beer swilling, whiskey drinking, individualistic cat. Grow your whiskers, get rid of fur. Be who you like. Be yourself."

"Yes, but," says the Cat, "then am I free to come and go just as I please?"

Alice shakes her head. She has decided it's her function to control the Cat.

"What about you?" the Cat inquires.

Alice shrugs, "Oh, I'm implicated."

While the Cat considers and reconsiders his position they turn to Mad Med.

"We gave you Virtue, Truth and Beauty."

"We gave you our memes."

"We gave you memories."

"Our past!"

"Your future!"

Med just stands there, and scowls at them. As they carry on, her scowl becomes more and more ferocious. The sisters stop. They ask Mad Med, "Well, what do you want?"

"I want–" Med hesitates. "I want to take over the building of Babel. I want my friends. I want to find out what I think of Babel. I want–" Here she grins an evil grin – "I want, if necessary, to destroy Babel."

The sisters brace themselves. If they had glass shields, they would now raise them.

"But you are ill-tutored."

"Ill-mannered."

"Illiterate."

Med glares.

"Your friends would trample on the memes of Babel."

Med shrugs.

"Babel wasn't built in a day, you know. Are you proposing the Conquest of Babel?"

Med turns away. "Keep Babel then. Keep it intact. May Crone Kronos and Death, and Mad Mem as well, haunt the ruins of Babel."

"What are you saying?"

"I am saying," Mad Med replies instantly, "that Babel cannot bloom in the desert air, and that a text doesn't exist until it's read."

The sisters bow their heads. "Very well," they say, "lower the barriers. Open the floodgates. Reveal the text. Let friends enter and barbarians as well. Are you content?"

Mad Med grins. "Come on," she cajoles, "It's not so bad."

The sisters tremble. Mad Med sets off to inform the world that Babel is now open to touts and tourists, vagrants and visitors, friends and allies, prospective immigrants, long lost citizens, and other pickers up of cultural artefacts. Very few come. Once again, the sisters tremble. They buckle down to help Mad Med.

# VII

## *THE READER'S TEXT*

We could rip off our masks. My own name is on the title page. And yours? We could meet, have a conversation, exchange messages. But the point, surely, is to exchange masks, not rip them off. You too have something to offer no doubt, and as long as it's entertaining and of some interest . . . There are things I'd like to know:

The Architectural Plans for Babel.

The Music to which the walls of Babel rose in the air.

Intelligent Conjecture on the true identity of The Black Piglet.

The Graffiti inscribed on the bricks of Babel.

What Skis or Motorbikes Sol and Shy used for surfing the sand dunes – a plausible design.

Bedtime Conversations between Cinders and the Prince.

The Contents of Verity's File of Intelligent Questions.

A Transcript of anything at all Charity ever said.

Altitude's Aspirations. (Who was Altitude? Her friendships and relationships and what finally became of her.)

**?**

if necessary, attach additional sheets.

Surprise me with something I didn't know I liked. On my part I promise to read anything readable. After all, that was the extent of your commitment. If it's good – my judgement – I am now exercising your prerogatives – I'll have a word with the publishers, get them to include it, try for payment [see invitation]. We're on the same side – more or less.

*"Who is the Stranger who breaks and distorts? Open isn't broken. The Reader is the stranger. Let him in. [Why 'him'?]"*

"We need immigrants to Babel. It's a necessity that Queen Alice never dealt with properly."

"Yes, builders for the buildings, skilled artisans, visitors to the site, hard working cultivators of memes and things."

"How attract them?"

"As flowers do, sister. As flowers attract visitors: with beauty and bribery".

"So Babel on the Internet? Babel on disk? How publish? How broadcast? How make the site known and seen?"

"By the use of Bulletin Boards, blank pages, wide margins, freeware and shareware, interactive modes – anything or something."

"But is this cultural imperialism on our part; and an encouragement to thievery and plagiarism on the part of babblers as yet unknown and unseen?"

"They won't plagiarise. Memes mutate. They will build new buildings."

"The domes of Babel will turn into towers, obelisks, all manner of obscenities. They will twist and crumble. All that we worked for turn into dead and forgotten memes. The Text is sacred. Do not distort it!"

"The text is alive. Text is the drapery and finery of Crone Kronos. It lives. It breathes."

"Poetic rubbish!"

"Look I'll show you. Here are some Bulletins."

"Where did you find them?"

"In the alleyways and back gardens, and just lying about on the streets."

"Rubbish!"

The Black Piglet sighs. Solitude sulks: scraps and tatters have been dumped in her lap. *[Laptop]* Some the wind carries away. The rest remain. She begins to read.

MAD MED'S MEDITATIONS WHEN SHE AND
ALICE WEREN'T ON SPEAKING TERMS

Multicultural memes. Like genes they mix. Like genes they need a host. A host to lodge in and to procreate. Genetic engineering. What corresponds? Can't put an animal meme into a different animal, or can you? Perhaps that's what we do when we domesticate animals? One way traffic? So far. Or misleading. So genes and memes have a common interest. Genocide kills the carriers of genes and culture. Extinct species and dead languages. Vast gene pools in danger of extermination. Keep them in a lab. till needed? Probably much better for a gene or a meme to find a live host. Not in books or in computers, but in brains. Does the brain have far more RAM than a computer? All that jumble instantly available? Colossal! Hindu society particularly has used live brains for ages and ages to store: Memorised epics, The gossip network, and Who was whose ancestor.

## MORALS

"Everyone's weak point their strong point. Heigh ho. Perhaps this leads to the Eightfold Path of Moderation," said The Black Piglet under the palm tree, the bodhi tree, in the arms of her Death. Why not. Who knows?

## ANONYMOUS TIPS

The Stuffit Expander needed for reading poetry. Not for prose. A difference in density, the only real difference. For example: Sometimes the memes lay thinly on the ground / And Alice ran around and around.

## IMPORTED MEMES

"And there's you and me my sweet duality." From *Sybil: The glide of her tongue*, Gillian Hanscombe.

"God has no pride." From *Temporary Shelter*, M. Travis Lane.

"Poor Child creeps through centuries of bone . . ." From *The Boatman*, Jay Macpherson.

"Anything worth doing is worth doing badly." Christine Donald in conversation.

Solitude frowns, and shuffles through the scraps she's been looking at. She needs an interruption. Reader approaches riding on horseback, camel back, mule back . . . A representative of great civilisations

186

put to the sword? Well, Reader will do. Let Reader have what Reader will.

"Who's in charge here?" Reader demands frowning at the litter lying in the sand.

"I am," Mad Med replies galloping up. "I know the answers to all these questions. Babel rose to the sound of saxophones. What's inscribed on the bricks of Babel don't really matter. They've been whitewashed. But what's plastered across the walls of Babel are outsize ads for my film fantasy. As for how and why Superman flies: he flies superbly with confidence and ease, as does the good ship *Enterprise*. And if you really want to know what Solitude thought about all alone in the desert, you'll find the answers in my best selling cookbook, *Saturnalian Secrets*. That's really obvious if you think about it. And as for morals, they're all worked out in my latest text on Interactive Logic. *[Fuzzy logic.]* Start with the theorem: 'Don't do unto others as you would have them do unto you. Respect their individuality.' And the rest follows – Hey!" Mad Med has seen the scraps in Solly's lap. *[Laptop]*

"You've been reading my diary!"

Meekly, Solly hands over the bits. She and The Black Piglet sneak away. They're tired. Let the younger generation deal with things. Meanwhile, Reader has been getting more and more cross.

"That's very aggressive," he tells Medusa.

"What is?"

"There's something indefinably aggressive about giving me a list of blank topics."

"But I was just filling them up for you."

"That's even worse."

"Okay, you fill them up." Mad Med is genuinely puzzled. "Building Babel is an ongoing project. Don't you see that 'the Reader's Text' is required by aesthetic logic?"

"The Aesthetic Logic is fine," Reader growls. He wants Med to know that he can deal quite easily with that sort of thing.

"Well?"

"What's wrong is the balance between the sprawl of Babel and the space left to me. How much blank white space do I have? Only a few lines? And confined to the margins? Why don't I just look for the next bit of print, the next set of memes, full grown and visible? If you really want me to contribute to Babel, then plead with me, and make it genuinely possible for me. We're fighting the convention of Reader as passive; we're fighting me, my need to go on and find out what happens. Tourists just gawp, they're not supposed to do anything."

But it's Mad Med who is gawping. "Okay, what do you want me to do?"

Reader frowns. "I want you to tell me to do something. Something specific."

"An interactive text?" asks Med uncertainly. "For instance: 'Now write down the main points. Do some exercises.' Like that?"

Reader scowls. "*Not* like that. I'm not a beanbag. Don't stuff my head full of memes and beans!"

"Well, what about: Here's a diagram. Fill in the missing bits. Or here's a Crossword half done, do the rest. Or tests with right answers. Riddles? Game Book Texts? For example: Did the Cat kill the Queen? If yes, go to page such and such. (I'm afraid you'll have to write that page.) Would that be all right?"

Reader frowns still more deeply. "I want something that's more like a game show – with prizes. After all, what's in it for me? You're probably getting paid for this."

"Who me? I'm just a figment of your imagination."

"Well, I'm not. I've got to eat."

"I'll ask the publisher," replies Med uncertainly. "It's the publisher who decides. It's the publisher who puts the Capital into Culture. That's how it is."

"What's in it for you?" asks Reader suddenly. "What do you want?"

"I want you to understand that you are part of the process of Building Babel. And I want you to do it, and I want you to enjoy it."

"You're awfully zealous about wanting me to have an entertaining time," Reader says slyly.

"I'm not zealous. Well, I am zealous. Listen, I'll come clean. I want you to become a writer, to understand that in the act of reading you are writing."

"What if I refuse? What happens if I say I'm just not playing?"

"What happens to your garden if you just leave it?"

"Weeds." replies Reader darkly.

"And what are weeds?" Med's patience is running out. There's an edge to her voice.

Reader grins. "Other people's memes!"

In spite of herself Medusa smiles; then she gets off her own high horse. With all the charm she can muster, she cajoles Reader, "Dear Reader, Sweet Barbarian, don't you understand that you *are* complicit? In the act of reading you inevitably build."

She helps him dismount, and gravely offers a handful of sand. He lets the sand slide through his fingers. It makes a pattern. Publishers and princes, piglets and parakeets, it all holds together. They will do something. In time they will all do something. With tentative sticks they begin to scratch and sketch in the sand: Crone Kronos' face – what it looks like – over and over. It's their quest, their occupation, something to do, a whodunit thriller, an ongoing opera, a myth, a story, a history . . . They sing a few songs. They do what people have always done. They write a few poems. They praise, they grieve.

QUEST, OCCUPATION, SOMETHING TO DO . . .

i)

> *What is most ancient is young.*
> *What most recent the stuff of the old.*
> *No mystery here, just any living thing,*
> *being brave, being bold.*

Once upon a time there was an old woman. Call her God, call her Crone Kronos, or call her The Enemy. Her problem was she was bored. She looked at a stone and the stone eroded. She looked at an atom and the atom exploded – or did not explode. It wasn't good enough. She wanted interaction, a voluntary response. And therefore she invented life. Thereafter consciousness. Thereafter language. And thereafter poetry, which made her so excessively cross she refused to speak, because then we reproached her and tried to say what cannot be said, and to tell the tale that may not be told.

ii)
In her garden a toad and a toadstool,
no bench for the aged to sit upon.
In her garden lies and lullabies,
this season's flowers, not yet shop-
                              soiled,
for the casual consumer to light upon.

In her garden a very old woman,
                blighting and blessing
whatever she chooses to chance upon.


iii)   *A Prayer*
Let me be your grand-child,
let me be you great grand-child
        – some descendant,
so that,
whatever the loss and dissolution
I am not, altogether, disinherited

iv)
Are you capable of kindness, saying,
"Sleep, baby, sleep?" Are you capable
of granting some extraordinary gift,
so that suddenly I'm able to fly
or to treat whole forests like so many
                toothpicks?
Yes, I know,
    you are capable of anything.

v)
Children should not rage, the puny should not
                      fight.
Who can know your innermost thought when
                    your most
horrid and eager dreams play out their birthright?

*If you would like to know more about Spinifex Press,
write for a free catalogue or visit our Home Page.*

**SPINIFEX PRESS**
PO Box 212, North Melbourne,
Victoria 3051, Australia
http://www.publishaust.net.au/~spinifex